Annals of Mathematics Studies

Number 127

Lectures on the Arithmetic Riemann-Roch Theorem

by

Gerd Faltings

Notes taken by Shouwu Zhang

PRINCETON UNIVERSITY PRESS

———

PRINCETON, NEW JERSEY
1992

The Annals of Mathematics Studies are edited by
Luis A. Caffarelli, John N. Mather, and Elias M. Stein

Princeton University Press books are printed on acid-free
paper and meet the guidelines for permanence and durability
of the Committee on Production Guidelines for Book
Longevity of the Council on Library Resources

Printed in the United States of America

10 9 8 7 6 5 4 3 2

Library of Congress Cataloging-in-Publication Data

Faltings, Gerd.
        Lectures on the arithmetic Riemann-Roch theorem / Gerd
Faltings; notes by Shouwu Zhang.
                p.   cm. — (Annals of mathematics studies ; no. 127)
                Includes bibliographical references.
                ISBN 0-691-08771-7 (CL)—ISBN 0-691-02544-4 (PB)
                1. Geometry, Algebraic.   2. Riemann-Roch theorems.
                I. Zhang, Shouwu.   II. Title.   III. Series.
        QA564.F36   1992
        516.3'5—dc20   91-42225

Table of Contents

In the spring of 1990 I gave a graduate course at Princeton University on the arithmetic Riemann-Roch theorem, which had just been shown by Bismut-Lebeau and Gillet-Soulé. The main purpose was to understand the techniques involved, and to simplify the presentation if possible. These notes arise from the course, however with some later improvements. Arakelov theory was invented with the purpose of applying techniques from algebraic geometry to arithmetic problems, especially to obtain a proof of the Mordell-conjecture. The main idea is to formulate algebraic properties at finite places in terms of metrics, and then to try to find analogues at infinity. In short one has to endow everyting with metrics. Around 1982 some progress was made. I showed that for arithmetic surfaces there is a Riemann-Roch theorem, and that one can use it to derive various analogues of properties of complex surfaces, for example the Hodge-Index theorem and the positivity of ω^2. The key to all this was the construction of volume forms on the cohomology of hermitian bundles. At the same time Quillen proposed a similar construction, using Ray-Singer analytic torsion.

This lead to new interest in this topic, and soon there was rapid progress: Deligne generalized the volume forms to more cases, and Gillet and Soulé developed an arithmetic intersection theory for general varieties, as well as hermitian K-theory. Then they joined efforts with Bismut and managed to define the determinant of cohomology of an arithmetic variety. One then could ask for a Riemann-Roch result for this determinant. It turned out that even for the projective space the immediate generalization of the classical Riemann-roch was false. To remedy this they introduced a secondary class $R(x)$ so that a modified version (using R) remains true. Bismut and Lebeau could prove a Riemann-Roch result for closed immersions, from which the Riemann-Roch for determinant bundles follows with some more work. The methods employed use stochastic integration, and are not easy to understand for anybody not familiar with this subject.

In these notes we replace all the probability-theory by considerations about heat kernels, thereby hopefully making the subject accessible to a wider audience. However I feel that it still remains complicated, and the reader should have a good general mathematical education. Due to my background I find algebraic and geometric arguments generally easier than analytic ones, so I tried to resort to the former as much as possible.

Now let us explain the contents:
In lecture one I give a proof of the classical Riemann-Roch theorem, and introduce various notation. Nothing at all is original. After that, in lecture

two, we introduce arithmetric Chern-classes. Here the method is new, more following the classical procedure of Grothendieck, but the results are all in [GS2].

In lecture three we study heat kernels. We avoid using too much technology, instead relying on a simple cutoff-procedure which works nicely in all instances we need it. We apply this in lecture four to study asymptotics. Especially we prove the index-theorem for the Dirac-operator. All the ideas are due to Bismut, Gillet, and Soulé and can be found in [BGS1]. Especially impressive is the computation of subdominant terms, which we reproduce at the end of lecture 4.

After that in lecture five we define direct images in arithmetic K- theory. This can be done only for smooth maps. Attempts to treat closed immersions have lead to enormous difficulties, and the final results are not as smooth as one would like them to be. Here our approach seems to be slightly original, although the main ideas can all be found in [BGS1]. Finally we prove a general Riemann-Roch in lecture six, of Grothendieck-type. We prove that there is a unique class $R(x)$ for which the Riemann-Roch is true. We do not attempt to compute it, except for the fact that it is odd. The method of proof uses deformation to the normal cone. To control the analytic singularities in this process we use a difference method, comparing to another deformation with the same singularities, but for which we already know the result.

Finally we conclude with a last lecture on the theorem of Bismut-Vasserot, estimating analytic torsion for ample line-bundles. Althrough this result is somehow disconnected from our main topic, the methods developed earlier work beautifully and thus it seemed reasonable to include it here. Shouwu Zhang has written the first two versions of these notes. I thank him very much for this difficult and demanding task. G. Kings and S. Mochizuki helped with constructive criticism, and S. Barbu did a great job in the final preparation of the manuscript.

The author was supported by NSF.

Lectures on the Arithmetic
Riemann-Roch Theorem

LECTURE 1. CLASSICAL RIEMANN–ROCH THEOREM

In this lecture we intend to prove the Riemann-Roch theorem for smooth morphisms of regular schemes. We define K-groups and Chow groups in the classical way. Then the Riemann-Roch for projective lines follows directly from the definitions. By the deformation of a regular embedding of a scheme to its normal cone, we reduce the problem to the case of projective bundles. Finally we prove the Riemann-Roch for projective bundles using flag schemes. We will follow this line when we prove the arithmetic Riemann-Roch theorem.

All schemes will be essentially of finite type over $\mathbb{Z}$. Reference: [BS].

K-Groups

Let X be a Noetherian scheme. Let $G(X)$ be the free abelian group generated by all coherent sheaves. Let $G'(X)$ be the subgroup of $G(X)$ generated by the elements of the form $[F_1] - [F_2] - [F_3]$ where F's form an exact sequence

$$0 \to F_2 \to F_1 \to F_3 \to 0.$$

The Grothendieck group $K'(X)$ of X is defined to be the quotient group of $G(X)$ by $G'(X)$. We get another group $K(X)$ by requiring that all the sheaves F are locally free. There is a morphism $K(X) \to K'(X)$.

Theorem 1.1. *If X is regular then $K(X) = K'(X)$.*

Proof. We prove the theorem by showing that every coherent sheaf has a finite resolution by vector bundles. Let $\{U_i : 1 \le i \le k\}$ be an affine covering of X such that $X - U_i$ is the support set of a Cartier divisor D_i. Let s_i be the section 1 of $O(D_i)$. Let F be a coherent sheaf on X. First of all we want to prove that F is a quotient of a locally free sheaf.

For each i we have that $F|U_i$ is generated by finitely many sections $\{e_{i,j} : 1 \le j \le n_i\}$. Since D_i is the support set of s_i there is a $m_i > 0$ such that $\{e_{i,j} s_i^{m_i}\}$ can be extended to sections of $F(m_i D_i)$ over X. We have a morphism

$$O(-m_i D_i)^{\oplus n_i} \to F$$

which is surjective over U_i. Combining all these morphisms for all i we obtain a surjective morphism

$$\oplus_1^k O(-m_i D_i)^{\oplus n_i} \to F.$$

3

This proves that F is a quotient of a vector bundle.

Let $n = \text{proj dim}(F)$. This is finite as X is regular. From the above argument we can find an exact sequence

$$O \to G \to E_n \to \ldots \to E_0 \to F \to O$$

where the E_i's are vector bundles, and G is locally free. This shows that the map $K(X) \to K'(X)$ is surjective and similarly one derives injectivity.

By the above theorem when X is regular, the group $K'(X)$ has a commutative ring structure such that $[E] \cdot [F] = [E \otimes F]$ for two vector bundles $[E]$ and $[F]$.

Equivalently $[E] \cdot [F] = \sum_{i=0}^{\infty} (-1)^i \cdot [Tor_i^{O_X}(E, F)]$ for arbitrary coherent sheaves E and F.

We want to explain some functorial properties for K-groups.

Let $f : X \to Y$ be a proper morphism. We define a push forward morphism f_* from $K'(X)$ to $K'(Y)$ by sending the class of a coherent sheaf F to $\sum (-1)^i [R^i f_* F]$. It is a consequence of the Leray spectral sequence that $(gf)_* = g_* f_*$ as morphisms of groups. So K is a covariant functor from the category of Noetherian schemes with proper morphisms to the category of abelian groups.

Theorem 1.2. *Let $\pi : P = \text{Proj}_X(E) \to X$ be a projective bundle over X of rank n. Then we have the following canonical isomorphisms*

(1) $f_*[O(i)] = [\text{Sym}^i E]$ *if $i \geq 0$.*

(2) $f_*[O(-i)] = 0$ *if $0 < i < n + 1$.*

Proof. The computation of Cech cohomology of $O(n)$ gives the result.

Let $f : X \to Y$ be a morphism of schemes. We define the pull back morphism f^* from $K(Y)$ to $K(X)$ by sending $[F]$ to $[f^* F]$ if F is locally free. f^* actually is a morphism of rings. In this way K becomes a contravariant functor from the category of regular schemes to the category of commutative rings.

In general K is not a covariant functor to the category of rings. But for any proper morphism $f : X \to Y$ we have $f_*[E \otimes f^* F] = f_*[E] \cdot [F]$ so $f_* K'(X)$ is an ideal of $K(Y)$.

Theorem 1.3. *Let X be a noetherian scheme. Then we have the following properties:*

(1) *Let $i : Y \to X$ be a closed subscheme of X. Denote by U the open scheme $X \setminus Y$ and $j : U \to X$ the structure morphism. Then we have the following exact sequence*

$$K'(Y) \xrightarrow{i_*} K'(X) \xrightarrow{j^*} K'(U) \to 0.$$

(2) Let $\pi : Y \to X$ be a affine bundle over X. Then $\pi^* : K'(X) \to K'(Y)$ is an isomorphism.

(3) Let $P = \text{Proj}_X(E) \to X$ be a projective bundle over X of rank n. Then the morphism $\oplus_{i=0}^n K'(X) \to K'(P)$ sending $(x_0, \cdots, x_n)$ to $\sum_{i=0}^n (\pi^* x_i) O(1)^i$ is an isomorphism.

Proof. (1) Firstly we prove that j^* is surjective. Let F be a coherent sheaf on U. Then F can be extended to X.

Now we prove that the $\ker(j^*) = \text{Im}(i_*)$. It is obvious that $j^* i_* = 0$. We need prove that the kernel of j^* is contained in the image of i_*. Let x be an element in $K(X)$ such that $j^* x = 0$. We want to prove that $x \in \text{Im}(i_*)$. As the extension from U to X is unique up to sheaves supported in Y, we may assume that $x = [F]$ with F supported in Y. Then F is a successive extension of sheaves annihilated by the ideal of Y, and thus $[F]$ is in the image of i_*.

(2) Injectivity follows by imbedding into a projective bundle and using (1), (3). So we need only prove that π^* is surjective. If $X' \subseteq X$ is closed and $U = X - X'$, then by (1) it suffices to prove the assertion for $Y \times_X U \to U$ and $Y \times_X X' \to X'$. By induction and a limit argument we reduce to the case where $X = \text{Spec}(A)$, A an artinian local ring, and $Y = X \times \mathbb{A}^n$. Using also induction over n we may assume that $n = 1$. Finally we replace A by its residue field k and remark that any $k[t]$-module has a finite free resolution.

(3) We first show that the map is surjective, i.e. $K'(P) = \sum_{i=0}^n \pi^*(K'(X)) \cdot O(1)^i$. By the previous technique we may assume that $X = \text{Spec}(k)$, k a field, and $P = \mathbb{P}_k^n$. We use induction in n, $n = 0$ being the trivial case. If $P_1 = \mathbb{P}_k^{n-1} \subseteq P$ denotes a hyperplane, $j : P_1 \hookrightarrow P$ the inclusion, then $j_*(O(1)^i) = O(1)^i - O(1)^{i-1}$. If x is an element of $K(P)$, then the restriction of x to $P - P_1 = \mathbb{A}_k^n$ is induced from X. If this restriction vanishes, then x is in $j_* K(P_1)$. Thus surjectivity follows. For injectivity: If $x = \sum_{i=0}^n \pi^*(x_i) O(1)^i = 0$, choose i maximal with $x_i \neq 0$. Then $x_i = \pi_*(x \cdot O(1)^{-i}) = 0$.

Contradiction.

CHOW GROUPS

Let X be a Noetherian scheme. Let p be a non-negative integer. Let $Z_p(X)$ be the free abelian group generated by all integral subvarieties of X of dimension p. The elements of $Z_p(X)$ are called p-cycles.

Let X' be a closed subscheme of X, let $X_1, \ldots, X_k$ the irreducible components of X and X_i' the maximal reduced subscheme of X_i. The local rings O_i of X_i at its generic points are of finite length, say m_i. The cycle $[X']$ is defined to the cycle $\sum_i m_i [X'_i]$.

Let $Z_p'(X)$ be the subgroup of $Z(X)$ generated by all $[\operatorname{div}(f)]$ for meromorphic functions f on a $p+1$-cycle of X. Then the p-th Chow group $A_p(X)$ is defined to be the quotient of $Z_p(X)$ by $Z_p'(X)$. Here, for a meromorphic function f on an irreducible $Y \subseteq X$, $\operatorname{div}(f)$ is a linear combination of $Z \subset Y$ of codimension one. The coefficient of Z can be computed as follows: Let $R = O_{Yz}$ denote the local ring of Y in the generic point z of Z. Then $f = g/h$, with $g, h \in R - \{0\}$, and the coefficient of Z is the difference of lengths $l(R/gR) - l(R/hR)$.

Let $f : X \to Y$ be a proper morphism. We can define the push forward $f_* : A_p(X) \to A_p(Y)$ as follows: Let Z an integral subvariety of X, then

$$f_*(Z) = \begin{cases} 0, & \text{if } \dim f(Z) < \dim(Z); \\ f(Z)[R(Z) : R(f(Z))], & \text{otherwise,} \end{cases}$$

where $R(Z)$ is the field of rational functions on Z. Via this A_* becomes a covariant functor from the category of regular schemes with proper morphisms to the category of abelian groups.

Let $f : X \to Y$ be a flat morphism. Then the pull back of cycles induces a pull back morphism of Chow groups. Such a property enables A_* to become a contravariant functor from the category of schemes with flat morphisms to the category of abelian groups.

For regular X the $A_*(X)$ have a ring structure by intersection theory. Here we just introduce an action on $A_*(X)$ by line bundles. Let Z be an integral subscheme of X and let L be a line bundle on X. If D denotes the Cartier divisor of a meromorphic section of L on Z then the class in $A_*(X)$ of $[D]$ does not depend on D. We write this class as $c_1(L) \cdot [Z]$. We have the following projection formula: Let L be a line bundle on X, let $f : X' \to X$ be a proper morphism and let α be a cycle on X'. Then

$$f_*(c_1(f^*L) \cdot \alpha) = c_1(L) \cdot f_*\alpha.$$

We have the following fundamental theorem for first Chern classes:

Theorem 1.4. *Let X be a scheme. Then $c_1(L) \cdot [Z]$ depends only on the class of $[Z]$ in $A_*(X)$. So $c_1(L)$ define a linear endomorphism on $A_*(X)$. Morever if M is another line bundle then $c_1(L) \cdot c_1(M) = c_1(M) \cdot c_1(L)$.*

Proof. First of all we claim that for any meromorphic section f of L and any meromorphic section g of M we have (in $A_*(Z)$)

$$c_1(L) \cdot [\operatorname{div}_Z g] = c_1(M) \cdot [\operatorname{div}_Z f].$$

By the projection formula we need only to prove the theorem after a proper morphism.

Let I_f be the ideal sheaf of O_Z consisting of all elements a such that af is a regular. Let $\pi : Z' \to Z$ be the blowup of Z along I_f and let E be the exceptional divisor. Then

$$\text{div}(\pi^* f) = C - E$$

for an effective Cartier divisor C on Z'. This allows us to reduce to the case that $\text{div}(f)$ is effective. Similarly we can reduce to the case that both $\text{div}(f)$ and $\text{div}(g)$ are effective.

We now assume that $\text{div} f$ and $\text{div} g$ are effective. Let $\pi : Z' \to Z$ be the blowup of Z along the ideal sheaf $fL^{-1} + gM^{-1}$ of O_Z. Denote by E the exceptional divisor of π. Then we have that

$$\pi^*(\text{div}(f)) = E + A, \ \pi^*(\text{div}(g)) = E + B$$

for effective Cartier divisor A, B on Z'. Define the excess intersection $\epsilon(\text{div} f, \text{div} g)$ by the formula

$$\epsilon(f, g) = \max\{\text{ord}_D(\text{div}(f)) \cdot \text{ord}_D(\text{div}(g)) | \ \text{codim}_Z D = 1\}.$$

It is easy to prove that A and B are disjoint and if

$$\epsilon = \epsilon(\text{div} f, \text{div} g) > 0$$

then $\epsilon(E, A) < \epsilon$ and $\epsilon(E, B) < \epsilon$. After several blowups we may assume that $\epsilon(\text{div} f, \text{div} g) = 0$, or $\text{div} f$ and $\text{div} g$ meet properly.

We now assume that $\text{div} f$ and $\text{div} g$ are effective and meet properly. Especially the restriction of f to $\text{div}(g)$ can be used to compute $c_1(L) \cdot \text{div}(g)$, and similar for $c_1(M) \cdot \text{div}(f)$. We write $\text{div} f = \sum_i m_i A_i$ and $\text{div} g = \sum_j n_j B_j$. Let C be a codimension two integral sub-scheme of Z. We want to prove that the multiplicity of C in $\sum_i m_i \text{div}_{A_i} g$ is equal to that in $\sum_j n_j \text{div}_{B_j} f$.

Locally near the generic point of C all line bundles are trivial, so f and g can be considered as section in $R = O_{Z,C}$. If one of f and g is not in the maximal ideal then the multiplicities above are both zero. So we may assume that f and g form a system of parameters of R.

Let $\text{Kos}(f, g; R)$ be the Koszul complex of R with respect to f, g:

$$0 \to R \xrightarrow{(f,g)} R^2 \xrightarrow{(g,-f)} R \to 0.$$

Let $\chi(\text{Kos}(f, g; R))$ is its Euler-Poincaré characteristic:

$$\chi(\text{Kos}(f, g; R)) = \sum_i (-1)^i \operatorname{length}_R(H_i(\text{Kos}(f, g; R))).$$

We reduce to prove that both multiplicities are equal to $\chi(\text{Kos}(f, g; R))$.

Let $\text{Kos}(g, R/fR)$ be the Koszul complex of R/fR with respect to g:

$$0 \to R/fR \xrightarrow{g} R/fR \to 0.$$

Define its Euler-Poincaré characteristic $\chi(\text{Kos}(g; R/fR))$ in the same way. It is easy to prove that

$$\chi(\text{Kos}(f, g; R)) = \chi(\text{Kos}(g; R/fR)).$$

Now $\chi(\text{Kos}(g; M))$ is additive in all R modules and vanishes if M is of finite length. Thus it depends only on the length of M at height one primes. If m_i denotes this length for $M = R/fR$ in a height one prime p_i (corresponding to A_i), we have

$$\chi(\text{Kos}(f, g; R)) = \sum m_i \operatorname{length}(R/(p_i + gR)).$$

This proves our claim.

Now the assertions of the theorem follow. Let W be an integral subscheme of X. Let $[\text{div } f]$ be a rational divisor on W then $c_1(L) \cdot [\text{div } f]$ is equal to $c_1(O_W) \cdot x$ for some x in $c_1(L) \cdot [W]$. So $c_1(L) \cdot [\text{div } f]$ is rational. This proves that $c_1(L) \cdot [Z]$ depends only on the class of $[Z]$ in $A_*(X)$. Also $c_1(L) \cdot c_1(M) \cdot [Z]$ and $c_1(M) \cdot c_1(L) \cdot [Z]$ are represented by the same cycle in $A_*(X)$ so they coincide.

Theorem 1.5. *Let X be a scheme. Then we have the following properties:*

(1) *Let $i : Y \to X$ be a closed subscheme of X. Denote U the open scheme $X \setminus Y$ and $j : U \to X$ the structure morphism. Then we have the following exact sequence*

$$A(Y) \xrightarrow{i_*} A(X) \xrightarrow{j^*} A(U) \to 0.$$

(2) *Let $\pi : Y \to X$ be an affine bundle over X of rank n. Then the map $\pi^* : A_k(X) = A_{k+n}(Y)$ is an isomorphism.*

(3) *Let $P = \operatorname{Proj}_X(E) \to X$ be a projective bundle over X of rank n. Then we have that the map $\oplus_0^n A_*(X) \to A_*(P)$ sending $(x_0, \cdots, x_n)$ to $\oplus_{i=0}^n \pi^*(x_i)c_1(O(1))^i$ is an isomorphism.*

Proof.

(1) The assertion is obvious.

(2) We first prove that π^* is surjective. Let U be an affine open set of X and $X' = X \setminus U$ the closed subscheme of X. By (1) we need only prove the assertion for $Y_U \to U$ and $Y_{X'} \to X'$. By induction on $\dim X$ we may assume that X is affine and $Y = X \times \mathbb{A}^n$. By induction on n may assume that $n = 1$ and Y is a trivial line bundle over X. Let Z be an integral subscheme of Y. We want to prove Z is rationally equivalent to the pullback of some cycle in X. We may assume the closure of $\pi(Z)$ is X and $Z \neq Y$. Let η be a generic point of X and Y_η be the generic fiber of Y over η. Then Z_η is a divisor of Y_η so it must be equivalent to 0. We can extend Z_η to a cycle Z' on Y wich is equivalent to 0. Now the closure of the image of support of $\pi(Z \setminus Z')$ is a proper subscheme of X. Injectivity will follow from (3).

(3) First of all we want to prove this map is a surjective morphism. By induction on $\dim(X)$ and (1) we may assume Y is affine and E is trivial. So we have a projective subbundle $i : P_1 \to P$ of P of rank $n-1$ such that $P \setminus P'$ is an affine bundle over X. The assertion follows from (2) and fact that $i_*(A(P_1)) = A(P) \cdot H$ $(H = c_1(O(1)))$.

The injectivity follows from the fact that $\pi(\pi^*(x) \cdot H^n) = x$ and $\pi_*(\pi^*(x) \cdot H^i) = 0$ for $0 \leq i < n$. We may assume that $x = [X]$ and X is a point. The assertion follows from the fact that $H^n \cdot [P]$ is represented by any section of P over X.

CHERN CLASSES

Let X be a regular scheme. Let E be a vector bundle of rank n. Denote by P the projective bundle $\mathrm{Proj}_X(\mathrm{Sym}\, E)$. We define the ith Chern class $c_i(E)$ of E as the operator $A_p(X) \to A_{p-i}(X)$ such that

$$H^n \cdot \pi^*(x) = \sum_{i=1}^{n} \pi^*((-1)^{i-1} c_i(E) \cdot x) \cdot H^{n-i}.$$

Let us write

$$c_t(E) = t^n - t^{n-1} c_1(E) + \dots$$

as a formal polynomial in t.

Theorem 1.6. *If*

$$0 \to E_1 \to E_2 \to E_3 \to 0$$

is exact then

$$c_t(E_2) = c_t(E_1) c_t(E_3).$$

Proof. Since everything commutes with base change so we may prove this formula after any base change which induces an injective map on Chow groups.

For a vector bundle E on X, let $\mathrm{Flag}(E)$ be the scheme which classifies the filtrations

$$0 = E^n \subset E^{n-1} \subset \ldots E^1 \subset E^0 = E$$

such that E^i/E^{i+1} are line bundles. The morphism $\mathrm{Flag}(E) \to X$ is the composition of a chain of projective bundles. By theorem 1.5 this morphism induces an injective map on Chow groups.

Then over $\mathrm{Flag}(E_1) \times \mathrm{Flag}(E_3)$ the bundles E_1 and E_3 have complete filtrations, so does E_2.

We claim that if E is a vector bundle over X and

$$0 = E^n \subset E^{n-1} \subset \ldots E^1 \subset E^0 = E$$

is a complete filtration for E then

$$c_t(E) = \prod_{i=0}^{n-1} c_t(E^i/E^{i+1}).$$

Obviously this implies our theorem.

The above equation is equivalent to

$$\prod_{i=0}^{n-1} c_1\big((E^i/E^{i+1})^{\otimes(-1)} \otimes O(1)\big) = 0.$$

We prove this equation by induction on n. $P' = P(E/E^{n-1})$ is the zero set of the map $E^{n-1} \to O(1)$. For any cycle α in P,

$$c_1(O(1) \otimes (E^{n-1})^{\otimes-1}) \cdot \alpha$$

represents an element β in $A_*(P')$. By induction we have

$$\prod_{i=0}^{n-1} c_1((E^i/E^{i+1})^{\otimes(-1)} \otimes O(1)) \cdot \alpha$$
$$= \prod_{i=0}^{n-2} c_1((E^i/E^{i+1})^{\otimes(-1)} \otimes O(1))|_{P'} \cdot \beta$$
$$= 0.$$

From the above theorem we have that c_t is a group homomorphism from $K(X)$ to $\mathrm{End}(A_*(X)[t])^*$. To do any computation with Chern-classes one uses the splitting principle to reduce to the case that the bundles have complete filtrations, so that in K-theory $E = \sum_i L_i$ with L_i line-bundles. Then any symmetric function in the $c_1(L)$ is a polynomial in the Chern-classes of E. For example the Chern character ch is defined as a ring homomorphism from $K(X)$ to $\mathrm{End}(A_*(X)) \otimes \mathbb{Q}$ such that $\mathrm{ch}(E)$ is a polynomial of $c_i(E)$ with coefficients in $\mathbb{Q}$ and

$$\mathrm{ch}(E) = \sum_i e^{c_1(L_i)} = \sum_{n=0}^{\infty} \frac{\sum_i c_1(L_i)^n}{n!}$$

if after a base change $E = \sum_i L_i$.

RIEMANN-ROCH THEOREM

The Todd class of E is defined as a polynomial of $c_i(E)$ such that

$$\mathrm{Td}(E) = \Pi \frac{c_1(L_i)}{1 - e^{-c_1(L_i)}}$$

if $E = \sum_i L_i$ after a base change.

Assume now $f : X \to Y$ be a projective and smooth morphism of relative dimension d. The main theorem of this lecture is

Theorem 1.7. *Assume X and Y are regular.*

$$f_*(\mathrm{ch}(E) \cdot \mathrm{Td}(T_{X/Y})) = \mathrm{ch}(f_* E)$$

in $A_*(Y)$.

Proof. We prove the theorem in several steps. We first remark that the assertion is compatible with composition, that is Riemann-Roch for (f, E) and $(g, f_*(E))$ implies it for (gf, E).

Riemann-Roch for a projective line bundle.

Let E be a rank two bundle on X. Riemann-Roch is valid for $f : X = \mathbb{P}(E) \to Y$.

By theorem 1.3 in this lecture we have that $K(X) \otimes \mathbb{Q}$ as a module over $K(Y) \otimes \mathbb{Q}$ is generated by O_X and $O(-1)$. We have $f_*O_X = O_Y$, and all other direct images vanish.

Let $h' = c_1(O(1)) - \frac{1}{2}c_1(E)$. From the exact sequence

$$0 \to O \to E^{\vee} \otimes O(1) \to T_{X/Y} \to 0$$

we have $T_{X/Y} \simeq \det(E)^{-1} \otimes O(2)$ and $c_1(T_{X/Y}) = 2h'$. Since

$$h' = \frac{h'}{1 - e^{-2h'}} - \frac{-h'}{1 - e^{2h'}},$$

so

$$\mathrm{Td}(T_{X/Y}) = \frac{2h'}{1 - e^{-2h'}} = h' + (1 + \text{ even powers of } h').$$

As we know $h'^2 = c_2(E) + \frac{1}{4}c_1(E)^2$ so $f_* h'^{2n} = 0$ for any $n \geq 0$ and $f_* h' = 1$. So finally

$$f_*(\mathrm{Td}(T_{X/Y})) = 1.$$

Also $f_*[O(-1)] = 0 = f_*(ch(O(-1)) \cdot \mathrm{Td}(T_{X/Y}))$, as $\frac{h' e^{-h'}}{1 - e^{-2h'}}$ contains only even powers of h'.

This completes the proof.

Deformation to the normal cone.

Deformation to normal the cone will be a very important tool in the study of Riemann-Roch.

Let $f : Y \to X$ be a proper smooth morphism and F be a coherent sheaf on Y. We introduce the following notations:

$$\mathrm{Err}_f(F) = \mathrm{Err}_Y(F) = f_*(ch(F) \cdot \mathrm{Td}_{Y/X}) - ch(f_* F).$$

Let $f : Y \to X$ be a smooth morphism and $Z \subset Y$ a closed embedding such that Z is smooth over X. Let $\tilde{Y}$ be the blow up of $Y \times \mathbb{P}^1$ along $Z \times \infty$. The fiber $\tilde{Y}_0$ of $\tilde{Y}$ over 0 is isomorphic to Y. The fiber $\tilde{Y}_\infty$ of $\tilde{Y}$ over ∞ is an union of the two schemes: an exeptional scheme $\tilde{Y}'_\infty$, which is isomorphic to $\mathbb{P}(N^* \oplus O_Z)$, and a scheme $\tilde{Y}''_\infty$, which is isomorphic to the blow up of Y in Z. Here $N^* = I_Z/I_Z^2$ denotes the conormal bundle.

Let $\tilde{i} : Z \times \mathbb{P}^1 \to \tilde{Y}$ be the canonical embedding. The morphism $\tilde{i}_0$ can be split into $i_0 : Z \to \tilde{Y}_0$ and $j_0 : \tilde{Y}_0 \to \tilde{Y}$. The morphism $\tilde{i}_\infty$ is split into $i_\infty : Z \to \tilde{Y}'_\infty$ and $j_\infty : \tilde{Y}'_\infty \to \tilde{Y}$.

We claim that

$$\mathrm{Err}_{\tilde{Y}_0}(i_{0*} F) = \mathrm{Err}_{\tilde{Y}'_\infty}(i_{\infty*} F)$$

for any coherent sheaf F on Z.

First we have

$$\mathrm{Err}_{\tilde{Y}_0}(i_{0*} F) - \mathrm{Err}_{\tilde{Y}'_\infty}(i_{\infty*} F)$$

$$= f_{0*}[(ch(i_{0*} F) \cdot \mathrm{Td}_{\tilde{Y}_0/X}] - f_{\infty*}[(ch(i_{\infty*} F) \cdot \mathrm{Td}_{\tilde{Y}_\infty/X}]$$

$$= \tilde{f}_*\{j_{0*}[(ch(i_{0*} F) \cdot \mathrm{Td}_{\tilde{Y}_0/X}] - j_{\infty*}[(ch(i_{\infty*} F) \cdot \mathrm{Td}_{\tilde{Y}_\infty/X}]\}$$

where $\tilde{f}$ is the structure morphism of $\tilde{Y}$ to X.

Secondly, let F' be the pull back of F to $Z \times P^1$ and let $\tilde{F}$ be the push forward of F' to $\tilde{Y}$. We have $i_{0*}F = j_0^*\tilde{F}$. Furthermore $ch(\tilde{F})$ is supported on $\tilde{\imath}(Z \times \mathbb{P}^1)$, and if $\tilde{T}$ denotes the logarithmic tangent-bundle of $\tilde{Y}/\mathbb{P}^1$ ($\tilde{T}$ = dual of $\Omega^1_{\tilde{Y}/X}(d\log \tilde{Y}_\infty)/\Omega^1_{\mathbb{P}^1}(d\log \infty)$), then near $i_0(Z)$ (respectively $i_\infty(Z)$) $\tilde{T}$ is isomorphic to T_Y (respectively $T_{\tilde{Y}'_\infty}$). It follows that the difference of the two errors is the image under $\tilde{f}_*$ of the product of $ch(\tilde{F}) \cdot Td(\tilde{T})$ and the divisor of the rational function $t: \tilde{X} \to \mathbb{P}^1$ (induced by second projection). It is thus rationally equivalent to zero.

Riemann-Roch for embeddings into projective bundles.

Let $g : Z \to X$ be a smooth morphism. Let E be a vector bundle of the form $O_Z \oplus N$. Denote by Y the projective bundle $P_Z(E)$ on Z. Let $i : Z \to Y$ be the embedding corresponding to the morphism $E \to O_Z$. Let $\pi : Y \to Z$ be the structure morphism of the projective bundle. Let $f = g\pi$ and consider Y as a scheme over X.

Suppose that Riemann-Roch is valid for the morphism π. Then for any coherent sheaf F on Z we have that

$$\mathrm{Err}_Z(F) = \mathrm{Err}_Y(i_*F).$$

By definition and because $\pi_*[O_Y] = [O_Z]$ we have

$$
\begin{aligned}
\mathrm{Err}_Y(i_*F) &= f_*(\mathrm{ch}(i_*F) \cdot \mathrm{Td}_{Y/X}) - \mathrm{ch}\, f_*i_*F \\
&= g_*\pi_*(\mathrm{ch}(i_*F) \cdot \mathrm{Td}_{Y/Z} \cdot \mathrm{Td}_{Z/X}) - \mathrm{ch}\, g_*F \\
&= g_*(\mathrm{ch}(\pi_*i_*F) \cdot \mathrm{Td}_{Z/X}) - \mathrm{ch}\, g_*F \\
&= g_*(\mathrm{ch}\, F \cdot \mathrm{Td}_{Z/X}) - \mathrm{ch}\, g_*F \\
&= \mathrm{Err}_Z\, F.
\end{aligned}
$$

Riemann-Roch for projective vector bundles.

Let $i : Z \to Y$ be a closed embedding of codimension 1 such that $g : Z \to X$ is smooth. Then for any coherent sheaf F on Z we have

$$\mathrm{Err}_Z(F) = \mathrm{Err}_Y(i_*F).$$

We have

$$\mathrm{Err}_Y(i_*F) = \mathrm{Err}_{Y_\infty}(i_{\infty*}F).$$

We thus may assume that $Y = P_Z(E)$, where $E = O_Z \oplus L$ be a split rank 2 bundle. The assertion follows from the previous.

Let E be a vector bundle on X of rank $n > 2$. We want to prove the Riemann-Roch for $P = P(E)$ over X. By a base change we may assume that E has a quotient bundle E' of rank n-1. $P' = P(E')$ is a Cartier divisor of $O_P(1)$. By theorem 1.3 in this lecture $K(P)$ as a $K(X)$ module is generated by $K(P')$ and $O_P(-1)$. By induction on the rank of E we need only to prove Riemann-Roch for $O(-1)$, provided Riemann-Roch is valid for all projective vector bundles of ranks less than n.

Let Y be the flag scheme which classifies the complete filtrations

$$0 = E_0 \subset E_1 \subset \cdots \subset E_n = E.$$

We have a morphism from Y to P which is a composition of projective bundles of ranks at most $n - 1$. E_1 is just the pull back of $O_P(-1)$. Since $f_*[O_{P_Z(F)}] = [O_Z]$ for any projective bundle $f : P_Z(F) \rightarrow Z$ the push forward of $[O_Y]$ is $[O_P]$. This implies that the push forward of E_1 is $O_P(-1)$. Since the Riemann-Roch is valid for the the morphism from Y to P, we need only to prove Riemann-Roch for E_1 and the morphism from Y to X.

Let Y' be the flag scheme which classifies the filtrations

$$0 = E_0 \subset E_2 \cdots \subset E_n = E$$

such that E_j is of rank j. Then Y is a projective line bundle over Y'. E is the canonical line bundle $O_Y(-1)$. Riemann-Roch is valid for $O_Y(-1)$ and the morphism from Y to Y' and gives 0 in both sides of the Riemann-Roch formula. Therefore the Riemann-Roch is valid for $O_Y(-1)$ and so for the morphism from Y to X.

Conclusion.

Now the Riemann-Roch for any projective smooth morphism $f : X \rightarrow Y$ follows easily. Let $\pi : P \rightarrow Y$ be a projective bundle of Y such that $f = \pi i$ with $i : X \rightarrow P$ a regular embedding. By deformation to the normal cone

$$\mathrm{Err}_X(F) = \mathrm{Err}_P(i_* F)$$

for any coherent sheaf on X. Riemann-Roch for P implies the assertion.

LECTURE 2. CHERN CLASSES OF
ARITHMETIC VECTOR BUNDLES

For a regular scheme X which is flat over $\mathbb{Z}$ and such that $X_{\mathbb{Q}} = X \times \mathbb{Q}$ is smooth over $\mathbb{Q}$ we define the arithmetic K-groups and the arithmetic Chow-groups and establish a theory of Chern classes. The arithmetic K-group is a refinement of the ordinary K-group: the arithmetic K-group is generated by all vector bundles which have a metric at the infinite place and all (p,p) type differential forms, such that any exact sequence of hermitian bundles gives the secondary Chern character form. The arithmetic Chow group is also a refinement of the ordinary Chow group: a cycle is an ordinary cycle plus a current at the infinite place. We can define the pull back and the push forward for arithmetic Chow-groups, and the pull back for arithmetic K-groups. The definition of push forward for arithmetic K-groups needs more tools. We define it and prove the Riemann-Roch in the next few lectures.

We use [GS1] and [GS2] as general references.

KÄHLER MANIFOLDS

Connections.

Let X be a real smooth manifold and let E be a vector bundle on X. We may think of E as a sheaf of groups over X and it is a C_X^∞ module over X, where C_X^∞ is the sheaf of all smooth fuctions over the open sets of X. A connection ∇ on E is a morphism of group sheaves

$$\nabla : E \to E \otimes \Omega^1$$

such that for any smooth function f on X and any smooth section e of E we have

$$\nabla(fe) = d(f)e + f\nabla(e).$$

Notice that ∇^2 is a morphism of bundles (i.e the morphism of C_X^∞ modules.)

$$E \to E \otimes \Omega^2$$

and gives a section of $\mathrm{End}(E) \otimes \Omega^2$. We call ∇^2 the curvature of E, for any connection ∇.

If E has an inner product $<,>: E \otimes E \to O_X$ we say a connection ∇ of E is Riemannian for $<,>$ if

$$< \nabla s_1, s_2 > + < s_1, \nabla s_2 > = d < s_1, s_2 >$$

where $<,>$ has been extended to an inner product on $E \otimes \Omega^*$ in the canonical way.

Now we suppose X is a complex manifold and E is a (holomorphic) vector bundle on X. We have a canonical morphism of group sheaves

$$\bar{\partial}: E \to E \otimes \Omega^{0,1}$$

such that for any smooth function f and any holomorphic section e of E we have $\bar{\partial}(fe) = \bar{\partial}(f)e$. We say a connection ∇ of E is holomorphic if the projection of ∇ to the space $E \otimes \Omega^{0,1}$ is $\bar{\partial}$.

Suppose now E has a hermitian inner-product metric $<,>: E \times E \to O_X$. A connection on E is said to be hermitian if

$$< \nabla s_1, s_2 > + < s_1, \nabla s_2 > = d < s_1, s_2 >$$

where $<,>$ has been extended to a hermitian inner product on $E \otimes \Omega_{\mathbb{C}}^*$ in the canonical way.

Notice that the hermitan metric on E induces a metric on $E \oplus \bar{E}$ and a connection on E also induces a connection on $E \oplus \bar{E}$. It is easy to see that a connection is hermitian on E if and only if the induced connection is Riemannian on $E \oplus \bar{E}$.

It is well known that for any hermitian holomorphic vector bundle E on a complex manifold X there is an unique holomorphic and hermitian connection ∇ of E and ∇^2 is a (1,1) form.

Kähler manifolds.

Let X be a real manifold with a metric. This means that we have an inner product on T_X. Then there is a unique Riemannian connection ∇^{LC} on T_X such that for any two vector fields x and y on X we have that

$$\nabla_x y - \nabla_y x - [x, y] = 0.$$

Actually such a connection can be defined by the formula

$$2 < \nabla_x y, z > = x < y, z > + y < x, z > - z < x, y >$$
$$+ < [x, y], z > + < [z, x], y > + < x, [z, y] > .$$

We call such a connection the Levi-Civita connection. It is well know that because ∇^{LC} has no torsion, the exterior derivative d on Ω_X^* is induced by ∇.

Let X be a complex manifold which has a hermitian metric. This means that we have a J-invariant hermitian inner product on T_X where J is the complex structure on X. We define a (1,1)-form by the formula

$$\omega(x, y) \;=\; \frac{i}{2\pi} \cdot <x, Jy> .$$

We say that the metric on X is Kähler if $d\omega = 0$ and we say X is Kähler if X has a Kähler metric.

Lemma 2.1. *The following conditions are equivalent:*

   (1) *The metric is Kähler .*

   (2) *For any $p \in X$, there is a holomorphic coordinate system $z_1, \cdots, z_n$ such that $g \equiv 1 \pmod{z^2}$, where $g = (g_{i,j})$ with*

$$g_{i,j} =< \frac{\partial}{\partial z_i}, \frac{\partial}{\partial z_j} > .$$

Proof. By definition we have

$$\frac{2\pi}{i} \cdot d\omega = -i \sum_{j,k,l} \{ \frac{\partial g_{j,k}}{\partial z_l} dz_l \wedge dz_j \wedge d\bar{z}_k + \frac{\partial g_{j,k}}{\partial \bar{z}_l} d\bar{z}_l \wedge dz_j \wedge d\bar{z}_k \}.$$

Thus the Kähler condition is equivalent to

$$\frac{\partial g_{j,k}}{\partial z_l} = \frac{\partial g_{l,k}}{\partial z_j}, \frac{\partial g_{j,k}}{\partial \bar{z}_l} = \frac{\partial g_{j,l}}{\partial \bar{z}_k}.$$

Assume (1). By a linear base change we may assume that $\{\frac{\partial}{\partial z_i}\}$ is an orthonormal basis for T_X at p. Replace z_i by

$$z'_j = z_j + \sum \frac{\partial g_{k,j}}{\partial z_l} z_k z_l$$

we will obtain that $g \equiv 1 \pmod{z^2}$. This is (2).

(2)$\rightarrow$(1) is clear

Theorem 2.2. *Let X be a hermitian manifold. Let ∇^H denote the unique holomorphic hermitian connection on T_X. Then the following conditions are equivalent:*

   (1) *X is Kähler .*

   (2) *$\nabla^{LC} = \nabla^H$.*

   (3) *$\nabla^{LC}(J) = 0$.*

   (4) *∇^H has no torsion.*

Proof. It is clear that (2) and (4) are equivalent and they imply (3).
    Assume (3). Then

$$\nabla_z^{LC}\omega(x,y) = z(\omega(x,y)) - \omega(\nabla_z^{LC}x,y) - \omega(x,\nabla_z^{LC}y)$$
$$= z < x, Jy > - < \nabla_z^{LC}x, Jy > - < x, J\nabla_z^{LC}y >$$
$$= 0,$$

since $\nabla^{LC}J = 0$ and ∇^{LC} is Riemannian. This implies that

$$d\omega(x,y,z) = \nabla_x\omega(y,z) - \nabla_y\omega(x,z) + \nabla_z\omega(x,y) = 0.$$

This proves that (3) implies (1).
    Now we assume (1). By the lemma, for any point p we can choose a
system of coordinates z such that $g \equiv 1 \pmod{z^2}$. If $\nabla^H \frac{\partial}{\partial z_i} = \sum \theta_{i,j} \frac{\partial}{\partial z_j}$
then we have

$$\theta = \partial g \cdot g^{-1} \equiv 0. \pmod{z}$$

So the torsion of ∇^H is zero. This proves that (1) implies (4).

Hodge theory.
    Let X be a real manifold. Denote by Ω^p the sheaf of complex smooth
differrential p-forms. Then we have the de Rham complex (Ω^*, d):

$$0 \to \Omega^0 \xrightarrow{d} \Omega^1 \xrightarrow{d} \Omega^2 \cdots .$$

This complex is acyclic and gives a resolution of the constant sheaf $\mathbb{C}$.
Take smooth sections of this complex. We obtain a complex $(C^\infty(\Omega^*), d)$.
Denote its cohomology by $H_{DR}^*(X, \mathbb{C})$ which agrees with ordinary singular
cohomology.
    Suppose X has a Riemannian metric. This induces a metric on Ω^* and
a pre-Hilbert structure on $C^\infty(\Omega^*)$. Let d^* be the adjoint of d. We define
the Laplacian Δ_X on $C^\infty(\Omega^*)$ by the formula

$$\Delta_d = dd^* + d^*d.$$

We call $\mathcal{H} = \ker \Delta_d$ the space of harmonic forms. It is obvious that ($\mathcal{H} = \ker d \cap \ker d^*$ and the canonical map $\mathcal{H} \to H_{DR}^*(X.\mathbb{C})$ is an isomorphism.
    If X is a compact Riemanian manifold, using the theory of the next
lecture, we can prove that $\mathcal{H}$ is finite dimensional and that the restriction
of Δ_d to the orthogonal complement of $\mathcal{H}$ has a smooth inverse G, the
Green's operator.
    Now assume that X is a complex manifold. Denote by $\Omega^{p,q}$ the sheaf of
smooth (p,q)-forms of X. For any non-negative integer p we have Dolbeault

complexes $(\Omega^{p,*}, \bar{\partial})$ and $(\Omega^{*,p}, \partial)$. Take smooth sections of these complexes and denote their cohomologies by $H_{\bar{\partial}}^{p,*}(X, \mathbb{C})$ and $H_{\partial}^{*,p}(X, \mathbb{C})$. They are isomorphic to $H^*(X, \Omega^p)$ and $H^*(X, \bar{\Omega}^p)$.

Suppose X is a hermitian manifold. As for the operator d we can define operators ∂^*, $\bar{\partial}^*$, Δ_∂ and $\Delta_{\bar{\partial}}$. We have harmonic spaces $\mathcal{H}_\partial$ and $\mathcal{H}_{\bar{\partial}}$ and isomorphisms to $H^*(X, \Omega^*)$ and $H^*(X, \bar{\Omega}^*)$. If X is a compact hermitian manifold, using the theory of the next lecture, we can prove that $\mathcal{H}_\partial$ and $\mathcal{H}_{\bar{\partial}}$ are finite dimensional and that the restrictions of $\Delta_{\bar{\partial}}$ and Δ_∂ to the orthogonal complements of $\mathcal{H}_{\bar{\partial}}$ and $\mathcal{H}_\partial$ have smooth inverses.

If X is a compact Kähler manifold one can prove that

$$\Delta_d = 2\Delta_{\bar{\partial}} = 2\Delta_\partial.$$

We have the Hodge decomposition theorem:

$$\mathcal{H}_\partial = \mathcal{H}_{\bar{\partial}} = \mathcal{H}.$$

K-GROUPS FOR ARITHMETIC VECTOR BUNDLES

Chern forms.

Let X be real manifold. Let E be a complex vector bundle on X with a connection ∇. We define the Chern polynomial c_t' as

$$det(t - \frac{i}{2\pi}\nabla^2)$$

and the Chern character form $\mathrm{ch}'(E)$ to be

$$\mathrm{tr}\exp(\frac{i}{2\pi}\nabla^2).$$

Notice that for any form ϕ with coeffients in $\mathrm{End}(E)$ we have $d(\mathrm{tr}\,\phi) = \mathrm{tr}[\nabla, \phi]$ so $d(\mathrm{tr}\,\nabla^{2n}) = 0$. In other words $c_t'(E)$ and $\mathrm{ch}'(E)$ are closed forms.

The following formulas are easy to prove:

(1)
$$c_t'(E_1 \oplus E_2) = c_t'(E_1) \cdot c_t'(E_2);$$

(2)
$$\mathrm{ch}'(E_1 \otimes E_2) = \mathrm{ch}'(E_1) \cdot \mathrm{ch}'(E_2);$$

(3) Let $f : Y \to X$ be a smooth morphism of complex manifolds then

$$c_t'(f^*E) = f^*c_t'(E).$$

Secondary Chern forms.

To define K-groups of hermitian vector bundles we need to study exact sequences of hermitian vector bundles.

Let X be a complex manifold of dimension n. Let p be a non-negative integer. Let $A'^p(X)$ (or $A'_{n-p}(X)$) be the space of all (p,p) forms and let $\tilde{A}^p(X)$ (or $\tilde{A}_{n-p}(X)$) be the quotient of $A'^p(X)$ modulo the sums of the images of ∂ and $\bar{\partial}$.

Theorem 2.3. *There is an unique way to attach to each exact sequence*

$$E : 0 \to E_1 \to E_2 \to E_3 \to 0$$

an element $\widetilde{\mathrm{ch}}(E)$ in $\tilde{A}^*(X)$ such that

(1)

$$\mathrm{ch}'(E_2) - \mathrm{ch}'(E_1) - \mathrm{ch}'(E_3) = \frac{\partial\bar{\partial}}{\pi i}\,\widetilde{\mathrm{ch}}(E);$$

(2) *For any map $f : X' \to X$ of complex manifolds*

$$\widetilde{\mathrm{ch}}(f^*E) = f^*\,\widetilde{\mathrm{ch}}(E);$$

(3) *When E is split in the sense of hermitian bundles $\widetilde{\mathrm{ch}}(E) = 0$.*

Proof. Let E be an exact sequence as above. Let us construct an exact sequence $\bar{E}$ on $\mathbb{P}^1 \times X$:

$$0 \to \bar{E}_1 \to \bar{E}_2 \to \bar{E}_3 \to 0$$

by $\bar{E}_1 = E_1(1)$, $\bar{E}_2 = (E_1(1) \oplus E_2)/E_1$ and $\bar{E}_3 = E_3$. We identify the E's with its pull back on $\mathbb{P}^1 \times X$, and take the embedding E_1 to $E_1(1)$ by the divisor ∞. Then the restriction of $\bar{E}$ to $0 \times X$ is E and that to $\infty \times X$ is split. Choose a metric on $\bar{E}$ which induces the given metric on $0 \times X$ and split at $\infty \times X$. We define

$$\widetilde{\mathrm{ch}}(E) = \int_{\mathbb{P}^1} [\mathrm{ch}'(\bar{E}_1) + \mathrm{ch}'(\bar{E}_3) - \mathrm{ch}'(\bar{E}_2)] \log |z|.$$

We need to show that the integral above does not depend on the choice of the metrics on $\bar{E}$. Suppose we have two metrics on $\bar{E}$ which have same value at $0 \times X$ and $\infty \times X$. As hermitian bundles we denote them by $\bar{E}'$ and $\bar{E}''$. Put some metric on $\bar{E}$ on $X \times \mathbb{P}^1 \times \mathbb{P}^1$ such that the restriction of this metric on $X \times \mathbb{P}^1 \times 0$ is $\bar{E}'$ and that on $X \times \mathbb{P}^1 \times \infty$ is $\bar{E}''$.

We use the same name for the differential operator $\frac{\partial\bar{\partial}}{\pi i}$ and its extension to distribution. It is easy to see

$$\frac{\partial\bar{\partial}}{\pi i}\log|w| = \delta_\infty - \delta_0.$$

So we have

$$\widetilde{ch}(E') - \widetilde{ch}(E'') = -\int_{\mathbb{P}^1}\widetilde{ch}(\bar{E})\frac{\partial\bar{\partial}}{\pi i}\log|w|$$

$$= \int_{\mathbb{P}^1\times\mathbb{P}^1}ch'(\bar{E}_2)\log|z|\frac{\partial\bar{\partial}}{\pi i}\log|w|$$

$$- \int_{\mathbb{P}^1\times\mathbb{P}^1}ch'(\bar{E}_1)\log|z|\frac{\partial\bar{\partial}}{\pi i}\log|w|$$

$$- \int_{\mathbb{P}^1\times\mathbb{P}^1}ch(\bar{E}_3)\log|z|\frac{\partial\bar{\partial}}{\pi i}\log|w|.$$

Since $ch'(\bar{E}_i)$'s are closed

$$\widetilde{ch}(E') - \widetilde{ch}(E'') = \int_{\mathbb{P}^1\times\mathbb{P}^1}ch'(\bar{E}_2)\frac{\partial\bar{\partial}}{\pi i}\log|z|\log|w|$$

$$- \int_{\mathbb{P}^1\times\mathbb{P}^1}ch'(\bar{E}_1)\frac{\partial\bar{\partial}}{\pi i}\log|z|\log|w|$$

$$- \int_{\mathbb{P}^1\times\mathbb{P}^1}ch'(\bar{E}_3)\frac{\partial\bar{\partial}}{\pi i}\log|z|\log|w|$$

$$= 0,$$

as $\int_{\mathbb{P}^1}\log|w| = 0$ and the rest of the integrand is constant on $\{z = 0\}$ (resp. $\{z = \infty\}$).

It is easy to check $\widetilde{ch}(E)$ satisfies (1),(2) and (3).

The same result as in the theorem holds if we replace ch' by any polynomial in the Chern-classes c_i'.

One can easily show various identities for $\widetilde{ch}(E)$, by deforming to the split case. For example if we tensor with an F, then $\widetilde{ch}(E\otimes F) = \widetilde{ch}(E)\cdot ch(F)$.

Example of a secondary class (without proof).

Assume we are given a hermitian vector bundle E on X, and change its metric. Then the above construction (applied to $id\colon E\to E$) gives a secondary class $\widetilde{ch}$ whose image under $\partial\bar{\partial}/\pi i$ is the difference of the two chern-characters. We study this infinitesimally:

Suppose we have a C^∞-family of hermitian metris $h(t)$ on E, inducing holomorphic connections $\nabla(t)$ and chern-characters.

$$ch'(E)(t) = tr(\exp(\frac{i}{2\pi}\nabla(t)^2)).$$

Then $\frac{\partial}{\partial t}(ch'(E)(t)) = \frac{\partial\bar{\partial}}{\pi i}\widetilde{ch}(E)(t)$.

We can compute $\widetilde{ch}(E)(t)$ as follows (for details see lecture 5): There exists an hermitian C^∞-Endomorphism $N(t)$ of E such that

$$\frac{\partial}{\partial t} < e_1, e_2 >_{h(t)} = < N(t)\, e_1, e_2 >_{h(t)}.$$

Then the derivative of $\nabla(\partial)$ (∂ a holomorphic vector-field) is easily computed to be

$$\frac{\partial}{\partial t}(\nabla(\partial)) = -[N(t),\, \nabla(\partial)] \qquad\qquad \text{(commutator)}$$

From this one calculates (which we shall do later) that
$\widetilde{ch}(E)(t) = tr(\frac{1}{2}N(t)\,\exp(\frac{i}{2\pi}\nabla^2)) = \frac{\partial}{\partial t_{t=0}} tr(\exp(\frac{i}{2\pi}\nabla^2 + \frac{1}{2}N(t)))$, i.e.
the right hand side has the right image under $\frac{\partial\bar{\partial}}{\pi i}$, and vanishes if $N(t) = 0$. A similar formula holds for other characteristic classes.

If for example we scale the metric on E by a factor λ^2, then the corresponding ch-class is

$$\log(\lambda) \cdot ch'(E)$$

Arithmetic K-groups.

Let X be a regular scheme which is flat over $\mathbb{Z}$. Then $X_{\mathbb{C}} = X \times \mathbb{C}$ is a smooth complex manifold.

Let F be the complex conjugate on $X_{\mathbb{C}}$. We fix the following notations. We denote by $A^{'*}(X)$, $A'_*(X)$ and $\widetilde{A}^*(X)$ the subspaces of $A^{'*}(X_{\mathbb{C}})$, $A^{'*}(X_{\mathbb{C}})$ and $\widetilde{A}^*(X_{\mathbb{C}})$ generated by all p-forms α^p such that $F^*\alpha^p = (-1)^p\alpha^p$. We denote by $\bar{G}(X)$ the free abelian group generated by all vector bundles which have a hermitian metric on $X_{\mathbb{C}}$ which is invariant under F.

Let $\hat{G}(X)$ be the direct sum of $\bar{G}(X)$ and $\widetilde{A}^*(X)$. Let $\hat{G}'(X)$ be the subgroup of $\hat{G}(X)$ generated by the elements with form $[E_2] - [E_1] - [E_3] - \widetilde{ch}(E)$ where

$$E : 0 \to E_1 \to E_2 \to E_3 \to 0$$

is an exact sequence. The Grothendieck group $\hat{K}(X)$ is defined to be the quotient of $\hat{G}(X)$ by $\hat{G}'(X)$.

The group $\hat{K}(X)$ has a commutative ring structure such that

$$[E_1] \cdot [E_2] = [E_1 \otimes E_2],$$

$$[E_1] \cdot [\omega_1] = [\mathrm{ch}'(E_1) \wedge \omega]$$

and

$$[\omega_1] \cdot [\omega_2] = [\frac{\partial\bar{\partial}}{\pi i} \omega_1 \wedge \omega_2],$$

where E_i are hermitian vector bundles and ω_i are differential forms.

Let $f : X \to Y$ be a morphism of schemes. We have a pullback morphism of rings $f^* : \hat{K}(Y) \to \hat{K}(X)$ by sending a hermitian vector bundle (resp. differential form) on Y to its pullback on X.

To define the push forward of $\hat{K}$ we need to study the index theorem. This will be done in the later lectures.

ARITHMETIC CHOW GROUPS

Arithmetic Chow groups.

Let X be a regular scheme of dimension $n + 1$ which is proper and flat over $\mathbb{Z}$. Then $X_{\mathbb{C}} = X \otimes \mathbb{C}$ is a smooth complex manifold of dimension n. Let F be the complex conjugation on $X_{\mathbb{C}}$. We denote by $\hat{Z}_p(X)$ the group of hermitian p-cycles which is the quotient group of the free abelian group generated by pairs (Z, g_Z) of an irreducible subvariety Z of X of dimension p and an (n-p, n-p) current g_Z on $X_{\mathbb{C}}$ such that

$$h_{Z_{\mathbb{C}}} = \delta_{Z_{\mathbb{C}}} - \frac{\partial\bar{\partial}}{\pi i} g_Z$$

is smooth, modulo the subgroup generated by $(0, \partial\alpha + \bar{\partial}\beta)$ for some currents α and β. Here $\delta_{Z_{\mathbb{C}}}$ denotes integration of $Z_{\mathbb{C}}$, which exists for example by resolution of singularities. Also Hodge-theory implies that any Z has a Green's current g_Z: Choose as $h_{Z_{\mathbb{C}}}$ the harmonic representative for $\delta_{Z_{\mathbb{C}}}$.

Let $\hat{Z}'_p(X)$ be the subgroup generated by the elements

$$(\mathrm{div}(f), -\log |f| \cdot \delta_Z)$$

where f is a meromorphic function on a subvariety Z of dimension $p - 1$ and $\delta_Z = \delta_{Z_{\mathbb{C}}}$. Then the arithmetic Chow group $\hat{A}_p(X)$ is defined to be the quotient of $\hat{Z}_p(X)$ by $\hat{Z}'_p(X)$.

We have the following maps:

$$\hat{A}_p(X) \to A_p(X) : (Z, g_Z) \to Z,$$

$$h : \hat{A}_p(X) \to A'_{p-1}(X) : (Z, g_Z) \to h_Z = \delta_Z - \frac{\partial\bar{\partial}}{\pi i} g_Z;$$

and

$$\epsilon : \tilde{A}_p(X) \to \hat{A}_p(X) : g \to \epsilon(g) = (0, g).$$

We call h the curvature morphism.

Later we shall prove some equalities in $\hat{A}_*(X)$. In general it will be easy to show that they hold after mapping to $A_*(X)$, or after h. Thus they hold up to $\epsilon(\mathcal{H}(X))$, where $\mathcal{H}(X) \subset \tilde{A}_*(X)$ denotes the kernel of $\partial\bar{\partial}$. At least this is true if X is compact. In this case $\mathcal{H}(X)$ is isomorphic to the cohomology of $X_{\mathbb{C}}$, since by Hodge theory of compact Kähler manifolds the kernel of $\frac{\partial\bar{\partial}}{\pi i}$ is harmonic up to the image of ∂ and $\bar{\partial}$. Note that in this case we may replace C^∞−forms by currents, without changing $\mathcal{H}(X)$.

We have an exact sequence

$$\tilde{A}_*(X) \xrightarrow{\epsilon} \hat{A}_*(X) \to A_*(X) \to 0.$$

The map ϵ need not to be injective. For example, if $\Gamma(O_X)$ has an invertible elment α which is not a root of the unity, then $\epsilon(0, \log|\alpha|) = 0$.

As for the functor A_* we can define the pushforward map and pullback map. The pullback morphisms of $\hat{A}_*$ can be defined for any smooth morphisms of scheme. The pushforward morphisms can be defined for any smooth, proper and surjective morphism.

First Chern classes.

The group $\hat{A}_*(X)$ has a commutative ring structure by intersection theory. We just want to introduce the action on $\hat{A}_*(X)$ by a hermitian line bundle. Let L be a hermitian line bundle on X. Let (Z, g_Z) be a cycle such that Z is irreducible. If s is a meromorphic section of L on Z then the class of

$$\hat{c}_1(L) \cdot (Z, g_Z) = (\mathrm{div}(s), -\log|s| \cdot \delta_Z + h_L g_Z)$$

in $\hat{A}_*(X)$ does not depend on the choice of s, where $h_L = \frac{\partial\bar{\partial}}{\pi i} \log|s| + \delta_{[\mathrm{div}\, s]}$ is the curvature form of L. Like theorem 1.3 have

Theorem 2.4. *Let X a arithmetic scheme. Then $\hat{c}_1(L) \cdot (Z, g_Z)$ depends only on the class of (Z, g_Z) in $\hat{A}_*(X)$. Morever if M is another hermitian line bundle then*

$$\hat{c}_1(L) \cdot \hat{c}_1(M) = \hat{c}_1(M) \cdot \hat{c}_1(L).$$

Proof. As in theorem 1.3 we need only to prove the following statement: Suppose L and M have holomorphic sections f and g over Z such that $A = \text{div}(f)$ and $B = \text{div}(g)$ are integral and have no common component, then

$$\hat{c}_1(L) \cdot (B, -\log|g|\delta_Z) = \hat{c}_1(M) \cdot (A, -\log|f|\delta_Z).$$

We need the following remarks:

The proof of theorem 1.3 implies that $\text{div}_A\, g = \text{div}_B\, f$ so we need only prove to that

$$(1) \qquad -\log|f| \cdot \delta_B - h_L \log|g|\delta_Z = -\log|g| \cdot \delta_A - h_M \log|f|\delta_Z.$$

If Z is smooth and the divisors have normal crossings we may assume that $X = Z$. The above equation follows from the definitions of the h_L and h_M and the equality

$$\frac{\partial\bar{\partial}}{\pi i}\log|f|\log|g| = \frac{\partial\bar{\partial}}{\pi i}\log|g|\log|f|. \quad (\text{modulo } \text{Im}(\partial) + \text{Im}(\bar{\partial}))$$

In general we use a resolution of singularities $\tilde{Z} \xrightarrow{\pi} Z \subset X$. Write $\pi^*(A) = A_1 + A_2$ and $\pi^*(B) = B_1 + B_2$, where A_1 and B_1 are coprime and $\pi_*(A_1) = A$ and $\pi_*(B_1) = B$, and $\pi_* A_2 = \pi_* B_2 = 0$. Also all divisors should have normal crossings. Put metrics on $O(A_1)$ and $O(B_1)$ and hence on $O(A_2)$ and $O(B_2)$. Now the assertion is easily shown to be independent of the choice of metrics, and equation (1) on Z is the pushforward of the following equation $(\tilde{1})$ on $\tilde{Z}$:

$$(\tilde{1}) \qquad -\log|f| \cdot \delta_{B_1} - h_L \cdot \log|g| \cdot \delta_{\tilde{Z}} = -\log|g| \cdot \delta_{A_1} - h_M \log|f|\delta_Z$$

By the previous calculation one shows that $(\tilde{1})$ holds up to integrations over components of A_2 or B_2, which vanish under π_*.

CHERN CLASSES

We want to define the Chern classes for a hermitian bundle E of rank n. Let $f : Y = \mathbb{P}(E)$ be the vector bundle associated to E. On Y we have a canonical exact sequence

$$\tilde{E} : 0 \to E' \to f^*E \to L = O(1) \to 0.$$

Let us put the quotient metric from f^*E on L and the subspace metric on E'.

Theorem 2.5. *Let $H = \hat{c}_1(O(1))$. Then*

    (1) *the map $\hat{A}_*(X)^{\oplus n}$ to $\hat{A}_*(\mathbb{P}(E))$ by sending $\{x_0, x_1, \ldots\}$ to $\sum_{i=0}^{n-1} x_i \cdot H^i$ is injective.*

    (2) *the image of the above map is the set of all objects whose curvature is of the form $\sum_i x_i' H'^i$.*

Proof. The theorem is valid for A_* via the forget morphism $\hat{A}_* \to A_*$, and for A_*' via the morphism h. So we only need prove the theorem for kernel of $\frac{\partial \bar{\partial}}{\pi i}$ in $\tilde{A}*$.

The first assertion follows from the fact that $\pi_*(\pi^* \alpha \wedge H'^{n-1}) = \alpha$ for all α, and $\pi_*(\pi^* \alpha \wedge H'^i) = 0$ for $0 \leq i < n - 1$.

The second assertion follows from the fact that if $\frac{\partial \bar{\partial}}{\pi i} \alpha = 0$ then α is closed up to the image of ∂ and $\bar{\partial}$. This implies that α is of form $\sum_i x_i' H'^i$ up to the image of ∂ and $\bar{\partial}$.

We define Chern classes $\hat{c}_i(E)$ as operators on $\hat{A}_*(X)$ such that if $\hat{c}_t(E) = t^n - t^{n-1}\hat{c}_1(E) + \cdots$ then

$$f^*\hat{c}_H(E) = \epsilon(\tilde{c}_H(\tilde{E})).$$

Here $\tilde{c}_H(\tilde{E})$ denotes the secondary characteristic class associated to the chern-polynomial, evaluated at H. This defines operators $\hat{c}_i(E)$ which induce $c_i(E)$ on $A^*(X)$ and $c_i'(E)$ on curvature.

This definition is natural if corollary 2.7 below should hold for the canonical exact sequence $\tilde{E}$ above.

Theorem 2.6. *Suppose $E = L_1 \oplus L_2 \oplus \ldots$ then*

$$\hat{c}_t(E) = \hat{c}_t(L_1) \cdot \hat{c}_t(L_2) \ldots.$$

Proof. We need to show that

$$\prod_{i=1}^{n} (\hat{c}_1(L) - \hat{c}_1(L_i)) = \epsilon(\tilde{c}_H(\tilde{E})).$$

We notice that $\hat{c}_1(L) - \hat{c}_1(L_i)$ is represented by $\hat{E}_i = (E_i, g_i)$ with $\cap_i E_i = \emptyset$ and $g_i = \log |s_i|$ for the morphism

$$s_i : L_i \to E \to L.$$

So the left hand side of the above equation is $\epsilon(\alpha)$ for a form α. Also the above equation can be shown in $A_*'(X)$ via h. Let $\alpha' = (\tilde{c}_H(\tilde{E}))$ and

$\beta = \alpha - \alpha'$ then $\frac{\partial \bar{\partial}}{\pi i} \beta = 0$. Now this is only a statement at ∞, so can be checked in closed forms. From now on we calculate in $\widetilde{A}_*(X_{\mathbb{C}})$.

We use induction on n to prove that $\beta = 0$.

Now

$$\hat{E}_1 \cdot \ldots \hat{E}_n = (0, i_* g(E_2 \cdot \ldots E_n|_{E_1}) + g_1 \wedge h_2 \wedge \cdots \wedge h_n)$$
$$= (0, i_*(\alpha_1) + g_1 \wedge h_2 \wedge \cdots \wedge h_n)$$

where $\alpha_1 = \prod_{i=2}^{n}(\hat{c}_1(L) - \hat{c}_1(L_i))|_{E_1}$ and α_1' has the analogeous meaning. If γ is any closed form then

$$f_*(\alpha' \wedge h_1 \wedge \gamma) = f_*(\alpha' \wedge (\delta_{E_1} - \frac{\partial \bar{\partial}}{\pi i} g_1) \wedge \gamma)$$
$$= (f|_{E_1})_*(\alpha'|E_1 \wedge \gamma) - f_*(\frac{\partial \bar{\partial}}{\pi i}\alpha' \wedge g_1 \wedge \gamma),$$

Since $\alpha'|_{E_1} = h_1 \cdot \alpha_1'$, and since $\alpha_1' = \alpha_1$ by the induction hypothesis

$$f_*(\alpha \wedge h_1 \wedge \gamma) = (f|_{E_1})_*(\alpha_1 \wedge h_1 \wedge \gamma) + f_*(h_1 \cdot \prod_{i=2}^{n} h_i \wedge g_1 \wedge \gamma)$$
$$= (f|_{E_1})_*(\alpha_1' \wedge h_1 \wedge \gamma) + f_*(h_1 \cdot \prod_{i=2}^{n} h_i \wedge g_1 \wedge \gamma)$$
$$= (f|_{E_1})_*(\alpha'|_{E_1} \wedge \gamma) + f_*(h_1 \cdot \prod_{i=2}^{n} h_i \wedge g_1 \wedge \gamma)$$
$$= f_*(\alpha' \wedge h_1 \wedge \gamma).$$

Hence $f_*(\beta \wedge h_1 \wedge \gamma) = 0$ for all closed forms γ and so $\beta \wedge h_1 = 0$. Symmetrically $\beta \wedge h_i = 0$ for all i.

Write $\beta = \sum_{i=0}^{n-1} \beta_i H^i$. If $\beta_{n-1} = 0$ then we derive that all $\beta_i = 0$ for $i \neq 1$. But β_{n-1} has degree 0 so it suffices to show that it vanishes at each point. We thus may assume that $X = \operatorname{Spec} \mathbb{C}$ and $E = \mathbb{C}^n$.

By embedding to $\mathbb{P}^1$ we may then assume that $X = \mathbb{P}^1$ and $L_1 = O(1)$ and $L_i = O$. Since $\beta h_1 = \beta h_2 = 0$ so $\beta(h_1 - h_2) = 0$. This implies that $\beta c_1(O_X(1)) = 0$ and this implies that $\beta_{n-1} = 0$.

Corollary 2.7. Let $E : 0 \to E_1 \to E_2 \to E_3 \to 0$ be an exact sequence of hermitian bundles. Then

$$\hat{c}_t(E_2) - \hat{c}_t(E_1)\hat{c}_t(E_3) = \tilde{c}_t(E).$$

Proof. The assertion is true if $E_2 = \oplus L_i$ and E_1 and E_3 are subsums. In general taking the base change

$$\mathrm{Flag}(E_1) \times \mathrm{Flag}(E_3) \to X$$

we may assume that E_1 and E_3 have complete filtrations and so does E_2. We now can construct an exact sequence

$$\bar{E} : 0 \to \bar{E}_1 \to \bar{E}_2 \to \bar{E}_3 \to 0$$

on $X \times \mathbb{P}^1$ such that over $X \times 0$ this is E and over $X \times \infty$ the bundle $\bar{E}_2$ is a direct sum of line bundles and $\bar{E}_1$ and $\bar{E}_3$ are subsums of $\bar{E}_2$. Now $\omega = \hat{c}_t(\bar{E}_2) - \hat{c}_t(\bar{E}_1)\hat{c}_t(\bar{E}_3) - \tilde{c}_t(\bar{E})$ has the property that $\frac{\partial\bar{\partial}}{\pi i}\,\omega = 0$ and ω is 0 over $X \times \infty$. This implies that $\omega = 0$ on $X \times 0$.

All the usual identities about Chern classes now holds. First of all by the result in lecture 1 and its anologue in forms they hold up to

$$\epsilon(\ker \partial\bar{\partial})/(\mathrm{Im}(\partial) + \mathrm{Im}(\bar{\partial})).$$

Then we can use the splitting principle and deformation to sums of line bundles.

LECTURE 3. LAPLACIANS AND HEAT KERNELS

In this lecture we intend to compute the asympototic expression for the diagonal values of the heat kernel of the Laplacian on Riemannian manifold $X \times \mathbb{R}^k$,where X is compact and the metric is invariant under translation on $\mathbb{R}^k$. First of all we study the Sobolev spaces and Garding's inequality and prove the existence of the smooth heat kernel. Then we prove that the asymptotic formula for the diagonal value $e^{-t\Delta}(x, x)$ depends only on local data when $t \to 0$.

SOBOLEV SPACES

Let $f \in C_0^\infty(\mathbb{R}^m)$. We define the Fourier transformation $\hat{f}$ of f to be

$$\hat{f}(\xi) = \pi^{-\frac{m}{2}} \int_{\mathbb{R}^m} e^{-ix\cdot\xi} f(x) dx$$

where if $x = (x_1, \cdots, x_m)$, $(\xi = \xi_1, \cdots, \xi_m)$ then $x\xi = x_1\xi_1 \cdots + x_m\xi_m$. It is easy to see that $|\hat{f}|_{L^2} = |f|_{L^2}$. So we can extend the Fourier transformation to an unitary endomorphism of $L^2(\mathbb{R}^m)$.

Let s be a real number and $f \in C_0^\infty(\mathbb{R}^m)$. We define

$$|f|_s^2 = \int (1 + |\xi|^2)^s |\hat{f}(\xi)|^2 d\xi.$$

The Sobolev space $H_s(\mathbb{R}^m)$ is defined to be the completion of $C_0^\infty(\mathbb{R}^m)$ with respect to the norm $\|\|_s$. For example the δ-distribution is in $H_s(\mathbb{R}^m)$ for $s < -\frac{m}{2}$ as its Fouriertransform is constant.

We fix the following situation :

By a triple (X, E, ∇_E) we mean a Riemannian manifold X of dimension m and a metrized vector bundle E and a Riemannian connection ∇_E on E. We almost always assume that X is a product $Y \times \mathbb{R}^k$ with Y compact, and that either E and ∇_E are the pullback of a metrized vector bundle E_Y and ∇_{E_Y} on $Y \times \mathbb{R}^k$, or at least a small perturbation of such. (See below for the definition of "small perturbation.") In fact what matters is that we can cover X by finitely many coordinate charts such that all transition functions have uniformly bounded derivatives, and that X has a complete Riemannian metric. The latter condition is necessary to construct cutoff-functions with small derivatives.

Let D_n be the open unit disc in $\mathbb{R}^n$. We can find a finite covering U_i of Y and open embeddings $\{\phi_i : U_i \to D_{m-k} \times \mathbb{R}^k\}$ such that the restrictions of E and ∇_E to U_i is the pullback of some pair $(E', \nabla_{E'})$ on D_{m-k}.

We have isomorphisms $\{\psi_i : E|_{U_i} \xrightarrow{\sim} O^d_{U_i}\}$ Let $\{g_i\}$ be a partition of the unity for $\{U_i\}$. If $e \in C^\infty_0(X, E)$ then $\psi_{i*}(g_i e)$ can be witten as $\phi^*(f_{i,1}, \cdots, f_{i,d})$ where $f_{i,j}$'s are smooth functions on R^m with compact support. Let s be a real number. We define

$$|e|_s = \sum_{i,j} |f_{i,j}|_s$$

for and $e \in C^\infty_0(X, E)$. It is not difficult to prove that the equivalence class of this norm on $C^\infty_0(X, E)$ does not depend on the choice of $\{U_i\}$, $\{\phi_i\}$, $\{\psi_i\}$ and $\{g_i\}$. The Sobolev spave $H_s(X, E)$ is defined to be the completion of $C^\infty_0(X, E)$ under this norm $\|_s$. One easily shows that $H_s(X, E)$ is the dual of $H_{-s}(X, E)$, using cutoff-functions on the $\mathbb{R}^k$-factor.

Let $\alpha = (\alpha_1, \cdots \alpha_m) \in \mathbb{N}^m$. We fix the following notations:

$$|\alpha| = \sum \alpha_i,$$

$$d^\alpha = \frac{\partial^{\alpha_1 + \cdots + \alpha_m}}{\partial x_1^{\alpha_1} \cdots \partial x_m^{\alpha_m}}.$$

Theorem 3.1.

(1) Let K be any compact set in X. Let $s > t$. If $\{f_n\}$ is an infinite sequence of functions in $C^\infty_K(X, E)$ which is bounded in the norm $\|_s$ then we can find a subsequence of $\{f_n\}$ which converges under the norm $\|_t$.

(2) Let t be a positive integer and $s > t + \frac{\dim X}{2}$. Then H_s is contained in $C^t(X, E)$ and we have a constant c such that

$$|e|_{\infty,t} \le c|e|_s$$

where $|e|_{\infty,t}$ is defined in a similar way as $|e|_s$: locally we define

$$|f|_{\infty,t} = \max_{x, |\alpha| \le t} |d^\alpha f(x)|.$$

(3) Let u, s, t, ϵ be real numbers such that $\epsilon > 0$ and $s > t$. Then we can find a constant c such that

$$|f|_t^2 \le \epsilon |f|_s^2 + c|f|_u^2.$$

Proof. Fix $\{U_i\}, \{\phi_i\}, \{\psi_i\}$ and $\{g_i\}$ as above. It is easy to see that we need only prove the theorem for U_i and for the trivial line bundle. We may assume that $X = \mathbb{R}^m$ and E is the trivial line bundle.

For (1) let $g \in C_0^\infty(\mathbb{R}^n)$ be a function such that $g = 1$ in a neighborhood of K. Then $g f_n = f_n$ and so

$$\hat{f}_n(x) = \hat{g} * \hat{f}_n(x) = \int \hat{f}_n(y) \hat{g}(x - y) dy.$$

Hence

$$|\hat{f}_n(x)|$$

$$\leq \int |\hat{g}(x - y)| |\hat{f}_n(y)| dy$$

$$\leq |f_n|_s [\int |\hat{g}(x - y)|^2 \cdot (1 + |y|^2)^{-s} dy]^{\frac{1}{2}}$$

$$\leq c \cdot h(x)$$

where h is a continuous function of x, and similar for all derivatives of $\hat{f}_n(x)$. So there is a subsequence of f_n which we will still label by f_n, such that $\hat{f}_n(x)$ converges uniformly on each compact set of $\mathbb{R}^n$. Let $s > t$ and r be a positive number, then

$$|f_j - f_k|_t^2$$

$$= \int |\hat{f}_j - \hat{f}_k|^2 (1 + |y|^2)^t dy$$

$$= \int_{|y| \leq r} |\hat{f}_j - \hat{f}_k|^2 (1 + |y|^2)^t dy + \int_{|y| \geq r} |\hat{f}_j - \hat{f}_k|^2 (1 + |y|^2)^t dy$$

$$\leq c_r |\hat{f}_j - \hat{f}_k|^2_{\max, D_r} + 2c(1 + r^2)^{t-s}.$$

We can fix a r so that the second term of the last line is arbitrarily small. For a fixed r we can choose bigger j, k so that the first term is arbitary small.

For (2) we notice that if $t = 0$ then

$$|f(x)|^2 = |\hat{\hat{f}}(x)|^2$$

$$= [\int e^{ix \cdot y} \hat{f}(y)(1 + |y|^2)^s (1 + |y|^2)^{-s} dy]^2$$

$$\leq c_1 |f|_s^2,$$

since $\int (1 + |y|^2)^{-s} dy$ is bounded for $s > \frac{m}{2}$. So we obtain that $|f|_{\infty, 0} \leq c_1 |f|_s$. For any α by the same method we can find c_α such that

$$|d^\alpha f|_{\infty,0} \le c_\alpha |f|_s.$$

Let $c_t = \max_{|\alpha| \le t}\{c_\alpha c_1\}$ then we have that

$$|f|_{\infty,t} \le c_t |f|_s.$$

The assertion in (3) follows imediately from the following estimate for all $x \in \mathbb{R}^m$:

$$(1 + |x|^2)^t \le \epsilon(1 + |x|^2)^s + c(1 + |x|^2)^u$$

where

$$c = \max_x (1 + |x|^2)^{s-u}((1 + |x|^2)^{t-s} - \epsilon).$$

ELLIPTIC DIFFERENTIAL OPERATORS

Let (X, E, ∇) be a triple as before. Let P be an endomorphism of group sheaf of E. We say that P is a differential operator if locally when $X = \mathbb{R}^m$ and $E = O^n$ then P can be written in the form

$$P = p(x, D) = \sum_{|\alpha| \le d} A_\alpha(x) D^\alpha$$

where $A_\alpha(x) \in \operatorname{End}_{O_X}(E)$ and $p(x, y) \in \operatorname{End}_{O_X}(E)(y)$. We alway assume that A_α are bounded in $\operatorname{End}_{O_X}(E)$. It is easy to check that P can be extend to be a bounded morphism from H_{s-d} to $H_s(X, E)$. We say that P is elliptic if locally

$$\det\left(\sum_{|\alpha|=d} A_\alpha \xi^\alpha\right) \ne 0$$

for $\xi \ne 0$.

We say P is uniformally elliptic if for a fixed data $\{U_i, \phi_i, \psi_i, g_i\}$ we can find a constant c such that

$$\left\|\sum_{|\alpha|=d} A_\alpha \xi^\alpha\right\| \ge c|\xi|^d$$

on U_i. P has constant coefficients if it is invariant under $\mathbb{R}^k$-translations in $X = Y \times \mathbb{R}^k$. We say that P is close to an operator P_0 with constant coefficients if $P = P_0 + \delta P$, where P_0 is constant and δP is small in the following sense:

- All derivatives of the coefficients of δP are uniformly bounded on X.
- The coefficients of the leading term (of degree d) have sup-norm $\leq \epsilon$, for a small

enough ϵ.

Usually assertions for constant P_0 generalize to such P, by treating δP as a perturbation. We have the following theorem:

Theorem 3.2. (Garding) *Let P be an uniformally elliptic differential operator on E such that P is close to an operator with constant coefficients. Then there exists a constant c such that*

$$|f|_{s+d} \leq c(|f|_s + |Pf|_s).$$

Proof. We show that $|f|_{s+d} \leq c(|f|_{s+d-1} + |Pf|_s)$. Using the trivialisations on U_i it suffices to treat the case where $X = \mathbb{R}^{m-k} \times \mathbb{R}^k$, and f has support in a product $K \times \mathbb{R}^k$, K a fixed compact. It suffices to show that any $x \epsilon K$ has a neighborhood U such that the assertion holds for f with support in $U \times R^k$. Also by perturbation we may assume that $P = P_0$ is $\mathbb{R}^k$ invariant. Let $x = 0$. We then have the following decomposition

$$P = P_0 + P_1 + P_2$$

where $P_0 = \sum_{|\alpha|=d} A_\alpha(0)D^\alpha$, $P_1 = \sum_{|\alpha|<d} A_\alpha D^\alpha$ and $P_2 = \sum_{|\alpha|=d}(A_\alpha - A_\alpha(0))D^\alpha$. We have

$$|Pf|_s \geq |P_0f|_s - |P_1f|_s - |P_2f|_s$$
$$= |\sum_{|\alpha|=d} A_\alpha(0)\xi^\alpha \hat{f}(\xi)|_s - |P_1f|_s - |P_2f|_s.$$

Let c_1 be the constant which does not dependent on p such that

$$|\sum_{|\alpha|=d} A_\alpha \xi^\alpha| \geq c_1 |\xi|^d.$$

For c_2 big enough we have

$$|Pf|_s \geq \frac{c_1}{2}|(1+|x|^2)^{\frac{d}{2}}\hat{f}|_s - c_2|f|_{s+d-1} - |P_2f|_s$$
$$= \frac{c_1}{2}|f|_{s+d} - c_2|f|_{s+d-1} - |P_2f|_s$$

It is obvious that $|P_2|$ continuously depends on the coefficients of P_2. But $P_2(0) = 0$. So there is a neighborhood U_p of p so that (with c_2 big)

$$|P_2f|_s \leq \frac{c_1}{4}|f|_{s+d} + c_2 \cdot |f|_{s+d-1}$$

for all $f \in C_0^\infty(U_p)$. Finally we have

$$|Pf|_s \geq \frac{c_1}{4}|f|_{s+d} - 2c_2|f|_{s+d-1}.$$

Quite similar to this proof one shows that if f and Pf are both in H_s, then f lies actually in H_{s+d}.

The following theorem is a beautiful application of theorem 3.2.

Theorem 3.3. *Let X be a compact Riemannian manifold. Let E be a hermitian vector bundle on X. Let $P : E \to E$ be an elliptic differential operator of order $d > 0$. Then $\ker P$ is of finite dimension.*

Proof. Let $f \in H_d$ with $|f|_0 = 1$ and $Pf = 0$. Then by theorem 3.2 $|f|_d \leq C$. Since $H_d \subset H_0$ is compact such $f's$ form a finite dimensional space.

It also follows that a formally self adjoint elliptic P defines a self adjoint operator in $L^2(X, E) = H_0(X, E)$. More precisely any P has a formal adjoint P^t such that
$< Pf, g >=< f, P^t g >$ for $f, g \epsilon C_0^\infty(X, E)$, Let $<,>$ denote the L^2-inner product defined by the metric of E and the measure on X given by its Riemannian metric. Then any pair (f_1, f_2) of elements of $L^2(X, E)$ which is perpendicular to $(-P^t g, g)$, all $g \epsilon C_0^\infty(X, E)$, is the L^2-limit of elements (f, Pf) with $f \epsilon C_0^\infty(X, E)$:
It follows that $f_2 = Pf_1$ in the sense of distributions, so f_1 is in $H_d(X, E)$. Approximate f_1 in the H_d-norm by elements $f \epsilon C_0^\infty(X, E)$.

Let us recall that an operator A in a Hilbert-space H is selfadjoint if its domain $D(A)$ is dense, if $< Ax, y > = < x, Ay >$ for $x, y \epsilon D(A)$, and if A coincides with its adjoint. If this holds we obtain a spectral measure E such that $A = \int \lambda \cdot E(d\lambda)$.

HEAT KERNELS

Let A be a closed, densely defined self-adjoint operator of a Hilbert space H such that for any $x \in$ domain (A) we have

$$< Ax, x >\geq -c|x|^2.$$

We denote by End H the set of bounded operators of H.

Lemma 3.4. *There is an unique smooth map* $K : (0, \infty) \to$ End H *such that*

$$\begin{cases} \frac{d}{dt} K(t) + AK(t) = 0 \\ \\ \lim_{t \to 0} K(t)x = x \text{ for any } x \, \epsilon H \end{cases}$$

Proof. We could use the spectral theorem. Another approach goes as follows. It has the advantage that it also works for perturbations $A + \delta A$ of a selfadjoint A, where δA satisfies $\|\delta A(x)\| \leq r\|A(x)\| + c\|x\|$, with constants r, c such that $r < 1$. ($x \in$ domain (A)): Define an arc

$$\gamma = \{z(t) = |t| - \epsilon - c - ti, t \in \mathbb{R}.\}$$

For any $\lambda \in \gamma$, $\alpha \in [-c, \infty)$ we have $|\lambda - \alpha| > \epsilon$. So we have that $\lambda - A$ has a continous inverse in End H and

$$\|(\lambda - A)^{-1}\| \leq \epsilon^{-1}.$$

We define

$$K(t) = e^{-At} = \frac{-1}{2\pi i} \int_{\gamma} e^{-\lambda t}(\lambda - A)^{-1} d\lambda.$$

Then it is easy to show that e^{-At} satisfies the conditions of the lemma.

Suppose now that two different operators K_1, K_2 satisfy the lemma. Then for some $x \neq 0$ we have $x(t) = (K_1 - K_2)x \neq 0$ so we have that

$$\begin{cases} \frac{d}{dt}x(t) + Ax(t) = 0 \\ x(0) = 0. \end{cases}$$

This implies that $\frac{d}{dt}|x(t)|^2 \leq 2c|x(t)|^2$. So

$$|x(t)|^2 \leq |x(0)|^2 e^{\frac{ct}{2}} = 0.$$

We get a contradiction.

Let (X, E, ∇) be a triple as before and P a positive self-adjoint differential operator on E. Then by the above lemma e^{-Pt} exists. We are interested to represent e^{-tP} by a smooth kernel function, the heat kernel.

Theorem 3.5. *Let P be positive selfadjoint operator on E which is uniformly elliptic and close to an operator with constant coefficients. Then e^{-Pt} is represented by a smooth kernel*

$$K(x, y, t) \in C^{\infty}(X \times X \times (0, \infty), \operatorname{End} E).$$

Proof. For any $n \geq 0$ and any $t > 0$ we have

$$t^n P^n e^{-tP} = \frac{-1}{2\pi i} \int_{\gamma} \lambda^n e^{-\lambda} (\lambda - tP)^{-1} d\lambda.$$

This implies that $t^n P^n e^{-tP}$ as an operator on L^2 has bounded norm uniform in t. By Garding's inequality we have that e^{-tP} maps L^2 to H_{∞}. So the morphism $f \to (e^{-tP} f)(x)$ is represented by an element $\alpha(x, t)$ in L^2. Let

$$K(x, y, t) = < \alpha(x, \frac{t}{2}), \alpha(y, \frac{t}{2}) > .$$

Then $K(x, y, t)$ is a smooth section of the hermitian vector bundle $\operatorname{End} E$ on $X \times X \times (0, \infty)$. It is easy to verify that $K(x, y, t)$ represents e^{-Pt}.

We need to understand how the heat kernels depend on the operators. We start from the non-homogeneous equation:

$$\begin{cases} \frac{d}{dt} x(t) + Ax(t) = y(t), \\ x(0) = x. \end{cases}$$

Such an equation has the unique solution

$$x(t) = e^{-At} x + \int_0^t e^{-sA} y(t - s) ds.$$

Now let δA be a self-adjoint densely defined operator on H such that there is $r \in (0, 1)$ and c with

$$| < \delta A x, x > | \leq r < Ax, x > + c < x, x > .$$

For any $x \in H$, let $x(t) = e^{-t(A + \delta A)} x$. Then $\delta A x(t)$ is defined and

$$\begin{cases} \frac{d}{dt} x(t) + Ax(t) = -\delta A x(t) \\ x(0) = x. \end{cases}$$

So

$$x(t) = e^{-At} x + \int_0^t e^{-sA} (-\delta A x(t - s)) dt$$

or

$$e^{-t(A + \delta A)} = e^{-tA} - \int_0^t e^{-sA} \delta A e^{-(t-s)(A + \delta A)} ds.$$

Repeat this several times to obtain

Lemma 3.6.

$$e^{-t(A+\delta A)}$$

$$= \sum_{k=0}^{n} (-1)^n \int\limits_{\substack{\alpha_i > 0 \\ \alpha_0 + \cdots \alpha_k = t}} e^{-\alpha_0 A} \delta A e^{-\alpha_1 A} \delta A \cdots e^{-\alpha_n A} d\alpha_1 d\alpha_2 \cdots d\alpha_n$$

$$+ R_{n+1}.$$

where

$$R_{n+1}$$

$$= \int\limits_{\substack{\alpha_i > 0 \\ \alpha_0 + \cdots \alpha_{n+1} = t}} e^{-\alpha_0 A} \delta A \cdots e^{-\alpha_n A} \delta A e^{-\alpha_{n+1}(A+\delta A)} d\alpha_1 \cdots d\alpha_n d\alpha_{n+1}.$$

This perturbation expansion will be very important for us. One consequence of it is the formula

$$\frac{d}{ds}\left(e^{-A(s)}\right) = - \int\limits_{\substack{a+b=1 \\ 0 \le a,b \le 1}} e^{-aA(s)} \frac{dA(s)}{ds} e^{-bA(s)} \, da.$$

LAPLACIAN FOR RIEMANNIAN VECTOR BUNDLES

Let (X, E, ∇_E) be a triple as before i.e. the metrics on E and X are close to product-metrics. We have an induced metric on $E \otimes \Omega_X = E \otimes T_X^\vee$. The L^2 norm

$$(f, g) = \int_X <f, g>$$

induces pre-Hilbert structures on $C_0^\infty(X, E)$ and $C_0^\infty(X, E \otimes \Omega_X)$. The connection ∇_E becomes a linear operator of pre-Hilbert spaces:

$$\nabla_E : C_0^\infty(X, E) \to C_0^\infty(X, E \otimes \Omega_X).$$

Lemma 3.7.

(1) ∇_E *has an adjoint* ∇_E^*.
(2) $\Delta_E = \nabla_E^* \cdot \nabla_E$ *is a positive, self-adjoint, and uniformly elliptic operator operator on* $C_0^\infty(X, E)$ *close to an operator with constant coefficients.*

Proof. The inner-product on E induces a pairing

$$<,>_E: (E \otimes \Omega^*) \otimes (E \otimes \Omega^*) \to \Omega^*$$

such that

$$< e_1 \otimes \omega_1, e_2 \otimes \omega_2 >_E = < e_1, e_2 > \omega_1 \wedge \omega_2.$$

If x is in $E \otimes \Omega^k$ then we have

$$d < x, y >_E = < \nabla x, y >_E + (-1)^k < x, \nabla y >_E .$$

Let dx be the volume form on X induced by the metric on X. We define an operator

$$* : \Omega \to \Omega^{m-1}$$

such that

$$< \omega_1, \omega_2 > dx = \omega_1 \wedge *\omega_2.$$

This operator induces a $*$ operator from $E \otimes \Omega$ to $E \otimes \Omega^{m-1}$. We define ∇_E^* by the formula

$$\nabla_E^*(\alpha) = -\frac{\nabla_E * \alpha}{dx}.$$

Let $\alpha \in C_0^\infty(X, E \otimes \Omega)$ and $\beta \in C_0^\infty(X, E)$ then we have

$$\begin{aligned}
(\beta, \nabla_E^* \alpha) &= \int_X < \beta, \nabla_E^* \alpha > dx \\
&= \int_X < \nabla_E \beta, *\alpha >_E \\
&= \int_X d < \beta, *\alpha >_E - \int_X < \beta, \nabla_E * \alpha >_E \\
&= \int_X < \beta, \nabla_E^* \alpha > dx.
\end{aligned}$$

This implies that ∇_E^* is the adjoint of ∇_E.

We want to compute Δ_E locally. Let p be a point of X. Let ∇^{LC} be the Levi-Civita connection on T_X. We may choose local coordinates $\{x_i\}$ for p such that $\{\frac{\partial}{\partial x_i}\}$ is an orthonormal basis for $T_X(p)$ and

$$\nabla_i^{LC} \frac{\partial}{\partial x_j} = \nabla_{\frac{\partial}{\partial x_i}}^{LC} \frac{\partial}{\partial x_j}(p) = 0.$$

We also choose a local ON-basis for E which has derivatives zero at p.

Now we have

$$\Delta_E(\alpha)(p) = -(\nabla_E * \sum \nabla_i \alpha \cdot dx_i)/dx$$
$$= -(\nabla_E \sum \nabla_i \alpha (-1)^{i-1} dx_1 \cdots \widehat{dx_i} \cdots dx_m)/dx$$
$$= -\sum \nabla_i^2 \alpha.$$

This proves that Δ_E is uniformly elliptic and positive. The fact Δ_E is symmetric on $C_0^\infty(X, E)$ follows from the definition.

Let $X = Y \times \mathbb{R}^k$ denote a Riemannian manifold as before, (E, ∇) a hermitian bundle with connection. We assume that all metrics are close to metrics induced from Y, so that all differential operators are close to constant ones. We consider such an operator P on E of the form $P = \Delta_E +$ terms of degree ≤ 1. Let $K(t) = K(t, x, y)$ denote the kernel representing e^{-tP}. We want to study the behavior of $K(t, x, x)$ as $t \to 0$. For example for $X = \mathbb{R}^m$ with the standard metric we have

$$K(t, x, y) = (4\pi t)^{-\frac{m}{2}} \cdot \exp\left(-\frac{(x-y)^2}{4t}\right)$$

Our strategy will be to compare $K(t, x, y)$ to this kernel, defined for the canonical metric on the tangent-space $T = T_{X,x}$ of X in x. So we choose a fixed $x \epsilon X$, and all estimates will be uniform in x.

Choose a small neighborhood U of x, and an open embedding $U \subseteq T$ which sends x to 0 and has differential one at x. In the following we consider a sequence of cutoff-functions φ_n which have support in an ϵ-neighborhood of x in U. We also assume that $\varphi_n = 1$ near x and on the support of φ_{n+1}. We can choose the φ_n's such that their k-th derivatives are bounded by a multiple (depending on n) of ϵ^{-k}. For the moment we choose ϵ fixed, but later we shall vary it proportional to t^α, some $\alpha > 0$, and thus we have to keep track of the dependance on ϵ of all constants. We assume that ϵ is bounded below by a multiple of $t^{\frac{1}{2}}$. We denote by Δ_n' the differential operator $\Delta_n' = \varphi_n P + (1 - \varphi_n)\Delta_T$ on T.

Let $K_n'(t, x, y)$ denote the kernel of $e^{-t\Delta_n'}$, defined for $x, y \epsilon T$. It follows that for $\ell \geq n$ $\varphi_\ell \cdot K_n'(t) \cdot \varphi_\ell$ and $\varphi_\ell \cdot K(t) \cdot \varphi_\ell$ can be considered as kernels on $U \times U$, and thus as operators on $L^2(T, E)$ as well as on Sobolev-spaces. However before we proceed we scale T by a factor of $t^{\frac{1}{2}}$. This way U now becomes a neighborhood of $0 \epsilon T$ of size $\sim t^{-\frac{1}{2}}$, tP becomes $P_t = \Delta_T + O(t^{\frac{1}{2}})$, and the heat-kernels $K(t), K_n'(t)$ get a factor $t^{\frac{m}{2}}$, because of the changes in volume-forms. Finally the cutoff-functions φ_ℓ now have k-th derivatives of size bounded by $\epsilon^{-1} t^{\frac{1}{2}} \ll 1$.

Now consider the operators (for $0 \leq u \leq 1$).
$$K_\ell(u) = \varphi_\ell \, e^{-uP_\ell} \, \varphi_\ell$$
and
$$K'_{n,\ell}(u) = \varphi_\ell \, e^{-u\Delta'_n} \varphi_\ell$$

Lemma 3.8. *Let $a \leq b$ denote integers*

1) *As operators from $H_a(T, E)$ to $(H_b(T, E)$ the norms of $K_\ell(u)$ and $K'_{n,\ell}(u)$ are bounded by a multiple of $u^{\frac{a-b}{2}}$.*

2) $\lim\limits_{u \to 0} K_\ell(u) = \lim\limits_{u \to 0} K_{n,\ell}(u) = \varphi_\ell^2$

3) $(\frac{\partial}{\partial u} + \Delta'_n)K_\ell(u) = [\Delta'_n, \varphi_\ell]e^{-uP_\ell}\varphi_\ell$
$(\frac{\partial}{\partial u} + \Delta'_n)K'_{n,\ell}(u) = [\Delta'_n, \varphi_\ell]e^{-u\Delta^1_n}\varphi_\ell$, *for $\ell > n$*

Proof. 1) follows easily from Garding's inequality, 2) is obvious, and so is 3) as on
$\mathrm{supp}\,(\varphi_\ell)$ Δ'_n and P_ℓ coincide.

Note that $[\Delta'_n, \varphi_\ell]$ is an operator of degree ≤ 1, whose coefficients have C^∞-norm $\ll \epsilon^{-1}t^{\frac{1}{2}} \ll 1$.

Theorem 3.9. *Assume $a \leq b$ are integers, n fixed.*

As an operator from $H_a(T, E)$ to $H_b(T, E)$ the norm of $\delta K_\ell(u) = K'_{n,\ell}(u) - K_\ell(u)$ is for $l > n+r$ bounded by a multiple of $\epsilon^{-r}t^{\frac{r}{2}}u^{\frac{a-b+r}{2}}$, for any integer r. The same bound holds for $\varphi_{\ell+1}K(u)\,(1 - \varphi_\ell)$ and $(1 - \varphi_\ell)\,K_\ell(u)\,\varphi_{\ell+1}$ (which express $K(u)$ away from the diagonal).

Proof. We show the assertion by induction on r. The case $r = 0$ follows 3.8. As $\delta K_\ell(t)$ approaches zero for $t \to 0$, and as

$$\left(\frac{\partial}{\partial u} + \Delta'_n\right)\delta K_{\ell+1}(u) = [\Delta'_n, \varphi_{\ell+1}]\delta K_\ell(u)\varphi_{\ell+1}$$

we have

$$\delta K_{\ell+1}(u) = \int\limits_0^u K'_n(u - v)[\Delta'_n, \varphi_{\ell+1}]\delta K_\ell(v)\varphi_{\ell+1} \cdot dv$$

Now estimate the norms in the integral as follows:
Note that $t^n\Delta^n e^{-t\Delta}$ has norm bounded by $\sup\{\lambda^n e^{-\lambda}|\lambda \geq 0\} < \infty$. Thus Garding's inequality implies bounds in Sobolov-norm for $e^{-t\Delta}$. The same is true for small enough perturbations.
If $0 \leq v \leq \frac{u}{2}$: $\delta K_\ell(V)\varphi_{\ell+1}$ as operator from H_a to H_a has norm bounded by $\epsilon^{-r} \cdot t^{\frac{r}{2}}u^{\frac{r}{2}}$.

$K'_n(u - v)$ as operator from H_{a-1} to H_b has norm bounded by $u^{\frac{a-b-1}{2}}$.
$[\Delta'_n, \varphi_{\ell+1}]$

as operator from H_a to H_{a-1} has norm bounded by $\epsilon^{-1}t^{\frac{1}{2}}$. So the integrand has norm $<< \epsilon^{-(r+1)}t^{\frac{r+1}{2}}u^{\frac{a-b+r-1}{2}}$.

If $\frac{t}{2} \leq x \leq t$:

$\delta K_\ell(S)$ as an operator from H_a to H_{b+1} has norm bounded by $\epsilon^{-r}t^{\frac{r}{2}}u^{\frac{a-b-1}{2}}$. K'_n $(t-s)$ as an operator from H_b to H_b has bounded norm. Finally $[\Delta'_n, \varphi_{\ell+1}]$ gives a contribution of size $\epsilon^{-1}t^{\frac{1}{2}}$.

All in all the integrand again has norm $<< \epsilon^{-(r+1)} \cdot t^{\frac{r+1}{2}} \cdot u^{\frac{a-b+r-1}{2}}$. By integrating the result follows.

The proof of the second assertion is quite analogous.

Corollary. $t^{\frac{m}{2}}\big(K(t,x,x,) - K'_n(t,x,x,)\big)$ *is* C^∞ *on* $[0,\infty) \times X$, *and has Taylor-series* $\equiv 0$ *at* $t = 0$.

This follows by integrating against $\delta_x \otimes \delta_x$ and use of the perturbation series in lemma 3.6: Only the behavior at $t = 0$ might cause problems. We have to show that $K(t,x,x)$ has an asymptotic expansion as $t \to 0$, and that the derivatives of this asymptotic expansion coincide with the perturbation series. But theorem 3.9 tells us that in this series we may neglect terms $K(t,y)$ where the distance of x and y is $>> t^{\frac{1}{4}}$, and in the remaining terms we may replace K by K'. A priori it only seems to be C^∞ in $t^{\frac{1}{2}}$, but it is invariant under $t^{\frac{1}{2}} \to -t^{\frac{1}{2}}$.

It remains to consider the Taylor-expansion of $t^{\frac{m}{2}} \cdot K'_n(t)$ at $t = 0$. For this we choose ϵ of size $t^{\frac{1}{4}}$ (any exponent between 0 and $\frac{1}{2}$ would do). Then after scaling $t\Delta'_n$ differs from Δ_T by an operator δP of order ≤ 2, whose coefficients have C^∞-norm bounded by a multiple of $t^{\frac{1}{4}}$. It follows that the Taylor-series in question is given by a perturbation expansion as in Lemma 3.6., i.e. by the sum

$$\sum_{\ell=0}^{\infty}(-1)^\ell \int_{\alpha_0+ \ +\alpha_\ell=u} e^{-\alpha_0\Delta_T} \cdot \delta P \cdot \ldots \cdot \delta P \cdot e^{-\alpha_\ell P} \, d\alpha_1 \cdots d\alpha_\ell.$$

Here the terms $e^{-\alpha_i\Delta_T}$ are classical heat- kernels, i.e.

$$e^{-\alpha_i\Delta_T}(x,y) = (4\pi\alpha_i)^{-\frac{m}{2}} e^{-\frac{|x-y|^2}{4\alpha_i}}.$$

As these decrease rapidly off the diagonal one then checks that in the formal expansion above we may replace δP by its Taylor-series at the origin (which has only finitely many terms modulo t^r, for any r). That is to compute the Taylor-series of $t^{\frac{m}{2}} \cdot K(t)$ we may treat P as a perturbation of Δ_T and expand everything formally in power-series. As a result we obtain

Theorem 3.10.

    a) $(4\pi t)^{\frac{m}{2}} K(t,x,x)$ *is* C^∞ *on* $[0,\infty) \times X$, *and has value 1 at* $t = 0$.

b) *Its Taylor-series at $t = 0$ can be computed as follows: Choose local coordinates at x and trivialize E. Expand P in a Taylor-series at x, and set $\delta P = P - \Delta_T$, where Δ_T is the constant Laplacian on $T = T_{x,x}$. Then formally expand $e^{-t(\Delta_T + \sigma P)}$ using the heat-kernel for $e^{-t\Delta_T}$ and the Taylor-expansion for δP.*

The resulting coefficients are polynomials in the coefficients of δP.

We also need a relative variant of the theory. Assume $f: X \to Y$ is a submersion of compact Riemannian manifolds, of dimensions m respectively k. Choose $x, y \in Y$, let $Z = f^{-1}(y) \subseteq X$, and $T = T_{Y,y}$. The resetriction of the tangent bundle T_X to Z splits orthogonally into $T_X Z = T_Z \oplus T_{Y,y}$. Choose an open neighborhood U of Z in X and embed it into $Z \times T$, with differential 1 on Z.

For a positive $s > 0$ consider the metric $g_s = g_X + s^{-1} f^* g_Y$ on X, and its Laplacian Δ_s. Its leading term looks like $\Delta_Z + s\Delta_Y +$ terms in s^2. We consider the heat kernel $e^{-t\Delta_s} = K(s,t)$ on X, and want to compare to $e^{-t(\Delta_Z + s\Delta_T)}$ on $Z \times T$. To do this we now use cutoffs φ_ℓ which are pullbacks of functions on Y, with support near y. However, as we have to use the metric g_s for Sobolev-norms, we change coordinates on $Z \times T$ by scaling the factor T by $\sqrt{s}$. In the new coordinates U has size $\sim 1/\sqrt{s}$ in the T-direction, and $\Delta_s = \Delta_Z + \Delta_T +$ lower order terms. We then first compare $K(s,t)$ to the kernel $K'(s,t)$ obtained by cutting of the difference to $\Delta_Z + \Delta_T$. For this one uses the differential equation (in t) satisfied by $\varphi_\ell (K(s,t) - K'(s,t)) \varphi_\ell$, and successive integration as before. The key is that the coefficients of $[\Delta_s, \varphi_\ell]$ have the norm bounded by $s^{\frac{1}{2}}$. All in all we see that $\varphi_\ell (K(s,t) - K'(s,t)) \varphi_\ell$ is $O((st)^r)$ for any r. Finally we compare $K'(s,t)$ to the heat-kernel for $\Delta_Z + \Delta_T$ which is the product of the heat-kernel on Z with the classical heat-kernel on T. All in all we easily obtain:

Theorem 3.11.

$$(4\pi t)^{\frac{m}{2}} \cdot s^{\frac{k}{2}} K(s, t, x, x) \text{ is } C^\infty \text{ on } [0, \infty)^2 \times X.$$

Its Taylor-series at $s = 0$ (respectively $s = t = 0$) can be computed by applying formally the perturbation-series of Lemma 3.6 to the Taylor-series, comparing to the heat-kernel on $Z \times T$ respectively $T_{Z,x} \times T = T_{X,x}$.

LECTURE 4. THE LOCAL INDEX
THEOREM FOR DIRAC OPERATORS

We will prove the local index theorem for Dirac operators on compact Kähler manifolds in this lecture. After some standard discussion on Dirac operators we prove the local index theorem in the absolute case by applying the results of the last lecture. The super-Dirac operators are defined to be the limits of Dirac operators in the sense of Clifford algebras.

CLIFFORD ALGEBRA

Let V be real vector space of dimension $2n$ with a quadratic form Q. The Clifford algebra $C(V) = C(V, Q)$ of V is defined to be

$$T(V)/ < v^2 + Q(v)|v \in V >$$

where $T(V)$ is the tensor algebra $\oplus_{r \geq 0} \otimes^r V$. (This is the analysts convention. In algebra one usually uses the opposite sign.) This is an algebra of degree 2^{2n} over $\mathbb{R}$ and there is a filtration F on $C(V)$ defined by

$$F_i(C(V)) = (\mathbb{R} + V)^i,$$

and $gr_F^* C(V) = \wedge^* V$. One example is the case $Q = 0$ which gives the exterior algebra $\wedge^* V$. We have a bijective map from $\wedge^* V$ to $C(V, Q)$ by sending $\lambda = v_1 \wedge \cdots \wedge v_l$ to

$$\lambda^a = \frac{1}{l!} \sum_\sigma \text{sgn}(\sigma) v_{\sigma(1)} \cdots v_{\sigma(l)}$$

where σ runs over the set of all permutation of $(1, 2, \cdots l)$. In this way we may think of Clifford algebras as deformations of the exterior algebras.

We assume that Q is non degenerate. Let $C(V) = C_+(V) \oplus C_-(V)$ be the decomposition of $C(V)$ into even and odd parts. We say elements in $C_+(V)$ have degree 0 and those in $C_-(V)$ have degree 1. We define the super-commutator $\{,\}$ in the following way:

$$\{x, y\} = xy - (-1)^{\text{degree } x \text{ degree } y} yx.$$

It is easy to see F_{2n-1} is generated by super-commutators.

Let us construct a complex representation for $C(V)$ as follows. Let J be a complex structure on V, that is an orthogonal endomorphism of V with the property that $J^2 = -1$. Let $V_{\mathbb{C}} = A \oplus A^*$ be the decomposition into eigenspaces of J. One obtains an action of $V_{\mathbb{C}}$ on $S(V) = \wedge^* A$: the elements of A operate by exterior multiplication and those in A^* by inner multiplication:

$$v^*(v_1 \wedge v_2 \wedge \cdots \wedge v_n) = -2 \cdot \sum_{i=1}^{n} (-1)^{i-1} < v^*, v_i > v_1 \wedge v_2 \wedge \widehat{v_i} \cdots \wedge v_n.$$

By counting dimensions we have that

$$C(V)_{\mathbb{C}} = \mathrm{End}_{\mathbb{C}}(S(V)).$$

Let $S = S_+ \oplus S_-$ be the decomposition into even and odd parts. Let ϵ be the endmorphism of S which is 1 in S_+ and -1 in S_-. The super-trace is then given by

$$\mathrm{tr}_s(x) = \mathrm{tr}\,\epsilon x.$$

Let v_i be a basis in A such that $< v_i, \bar{v}_j >= \delta_{i,j}$, and let $\omega = v_1 \wedge \bar{v}_1 \wedge \cdots v_n \wedge \bar{v}_n$. Then $(-\frac{1}{2})^n \omega\, \mathrm{tr}_s$ is just the morphism

$$\mathrm{Tr}_s : C(V) \to C(V)/F_{2n-1} = \wedge^{2n}(V).$$

Now we want to study the limit of Clifford algebras. Let Q_n be a series of quadratic forms on V. We say a series of Clifford elements $c_n \in C(V, Q_n)$ converges if there is a series of elements t_n in $\wedge^*(V)$ which converges and whose elements have images c_n in $C(V, Q_n)$. It is easy to see that if c_n converges then $\mathrm{Tr}_s(c_n)$ converges to $\mathrm{Tr}_s\, t_0$, where t_0 is the limit of the t_n.

We assume now that Q_n has a limit Q_0. Then if c_n converges then we have t_0 in $T(V)$ as above whose image c_0 in $C(V, Q_0)$ will not depend on the choice of t_n. We call c_0 the limit of c_n. In this case we have

$$\lim(c_n + c_n') = \lim c_n + \lim c_n'$$

and

$$\lim c_n c_n' = \lim c_n \lim c_n'.$$

Dirac operators

Let X be a Kähler manifold of dimension n. Let $\Omega_{X/\mathbb{R}}$ be the the bundle of one forms with induced metric. Let

$$\Omega_{X/\mathbb{C}} = \Omega_{X/\mathbb{R}} \otimes_{\mathbb{R}} \mathbb{C} = \Omega^{1,0} \oplus \Omega^{0,1}$$

be the decomposition into holomorphic and anti-holomorphic parts according to the complex structure of X. We denote by $C(X)$ the Clifford algebra with respect to the inner product on $\Omega_{X/\mathbb{R}}$. Let $S(X) = \bar{\Omega}^* = \wedge^* \Omega^{0,1}$ be the bundle of anti-holomorphic forms. Then we have a complex representation of $C(X)$ on $S(X)$. As metric on $S(X)$ we choose on $\Omega^{0,p}$ 2^p-times the hermitian metric. In this way Clifford-multiplication by real elements of Ω^1_X is antihermitian.

In the following we shall distinguish between $\bar{\Omega}^*$ and $S(X)$ as they have different metrics. With these conventions $S(X)$ is well adapted to the study of the Dirac-operator.

Let E be a hermitian vector bundle on X with the unique hermitian and holomorphic connection ∇_E. Let ∇^{LC} be the Levi-Civita connection on T_X which induces a connection on $S(X)$ which we still denote ∇^{LC}. Let $\nabla^{LC} \otimes 1 + 1 \otimes \nabla_E$ be the induced connection on $S \otimes E$ which is hermitian and holomorphic. Locally let e_i be a basis for T_X and let f_i be the dual basis for Ω_X. We have the following definition

Definition: *The Dirac operator D associated to ∇_E: $S \otimes E \to S \otimes E$ is defined to be*

$$D = \sum_{i=1}^{2n} f_i \nabla_{e_i}.$$

This is independant of the choices.

We have the following properties for Dirac operators:

Theorem 4.1:

(1) *Let D^* be the adjoint of D with respect to the L^2 norm. Then*

$$D^* = D.$$

(2) *Let $\bar{\partial}^*$ be the adjoint of $\bar{\partial}$ (for the metric on $S(X)$). Then*

$$D = \bar{\partial} + \bar{\partial}^*.$$

Proof: Let x be a vector field on X. Then we have a diffeomorphism for any small $t > 0$:

$$\exp_t(x) : X \to X$$

by sending (p, t) to $\exp_p(tx(p))$. This induces an endomorphism of $\wedge^* T_X$: $y \to \exp(tx)^* y$. We denote L_x the differential of this morphism:

$$L_x(y) = \frac{d}{dt} \exp(tx)^* y_{t=0}.$$

It is easy to see that L_x is a derivation on $\wedge^* T_X$, $L_x(f) = x(f)$ if f is a function, and $L_x(y) = [x, y]$ if y is a vector field. In the same way we can define a derivation L_x on $\wedge^* \Omega_X$:

$$L_x(\omega) = \frac{d}{dt} \exp(tx)^* \omega.$$

We have that for any y in T_X^* and any ω in Ω^*

$$x(\omega, y) = (L_x \omega, y) + (\omega, L_x y),$$

and also

$$\int_X L_x(< f, g > dx) = 0.$$

In other words

$$\int (< \nabla_x f, g > + < f, \nabla_x g > + \frac{L_x(dx)}{dx} < f, g >) dx = 0.$$

This implies that

$$\nabla_x^* = -\nabla_x - \frac{L_x(dx)}{dx}.$$

Assume that the e_i are orthonormal then $e_1 \wedge \cdots e_{2n}$ is dual to dx. So

$$-\frac{L_x dx}{dx} = \frac{L_x(e_1 \wedge \cdots \wedge e_{2n})}{e_1 \wedge \cdots \wedge e_{2n}}$$

$$= \sum_{i=1}^{2n} < [x, e_i], e_i >$$

$$= \sum_{i=1}^{2n} < \nabla_x e_i - \nabla_{e_i} x, e_i >$$

$$= -\sum_{i=1}^{2n} < \nabla_{e_i} x, e_i > .$$

$\left(\text{as} < \nabla_x e_i, e_i >= \frac{1}{2} x < e_i, e_i >= 0 \right)$ Finally

$$D^* = -\sum_i \nabla^*_{e_i} f_i$$

$$= \sum \nabla_i f_i + \sum_{i,j} < \nabla_j f_i, f_j > f_i$$

$$= \sum_i \nabla_i \cdot f_i - \sum_j \nabla_j f_j$$

$$= \sum_i f_i \nabla_i.$$

This proves (1). For (2) we may choose a complex basis e_i for $T_X^{1,0}$ and f_i its dual in $\Omega_X^{1,0}$. Then

$$D = \sum f_i \nabla_{e_i} + \sum \bar{f}_i \nabla_{\bar{e}_i}.$$

But $\nabla_{\bar{e}_i} = \bar{e}_i$ so $\sum \bar{f}_i \nabla_{\bar{e}_i} = \bar{\partial}$. (1) is equivalent to

$$\left(\sum f_i \nabla_{e_i} \right)^* + \bar{\partial}^* = \left(\sum f_i \nabla_{e_i} \right) + \bar{\partial}.$$

Comparing degrees we obtain that

$$\sum f_i \nabla_{e_i} = \bar{\partial}^*.$$

This proves (2).

Recall that the curvature of a connection ∇ is ∇^2. It is not difficult to prove that

$$R_{i,j} = (\nabla^2, e_i \wedge e_j) = \nabla_i \nabla_j - \nabla_j \nabla_i - \nabla_{[e_i,e_j]}.$$

Theorem 4.2: (Lichnerowicz) *Let Δ be the Laplacian of the vector bundle $S \otimes E$ with connection ∇. Then*

$$D^2 = \Delta - \frac{1}{4} \sum_{i,j} < R_{i,j}(f_i), f_j > + \sum_{i<j} f_i f_j \otimes \left(R_{i,j}(E) + \frac{1}{2} \, \text{tr}(R_{i,j}(T_X^{1,0})) \right)$$

Proof: Let p in X. We may choose an orthonormal basis of T_X such that $\nabla_i e_j = 0$ at p. Then at p we have

$$
\begin{aligned}
D^2 &= \sum_{i,j} f_i \nabla_i f_j \nabla_j \\
&= \sum_{i,j} f_i (\nabla_i f_j) \nabla_j + \sum_{i,j} f_i f_j \nabla_i \nabla_j \\
&= \sum_{i,j} f_i f_j \nabla_i \nabla_j \\
&= -\sum_i \nabla_i^2 + \sum_{i \neq j} f_i f_j \nabla_i \nabla_j \\
&= -\sum_i \nabla_i^2 + \sum_{i<j} f_i f_j (\nabla_i \nabla_j - \nabla_j \nabla_i) \\
&= -\sum_i \nabla_i^2 + \sum_{i<j} f_i f_j \otimes R_{i,j}(E) + \sum_{i<j} f_i f_j R_{i,j}(X) \otimes 1.
\end{aligned}
$$

$R_{i,j}(X) = R$ acts as a derivation on $S = \wedge^* \Omega^{0,1}$. Change notations for a moment and denote by e_i a basis of the holomorphic differentials and f_i the dual basis of antiholomorphic differentials, under the complex linear extension of the metric to $\Omega_{X,\mathbb{C}} = \Omega_x^{1,0} \oplus \Omega_x^{0,1}$. It then follows easily that on S

$$
\begin{aligned}
R &= \frac{-1}{2} \sum_{i,j} < R(f_j), e_i > f_i\, e_j \\
&= \frac{1}{2} \sum_{i,j} < R(f_j), e_i > (e_j \wedge f_i)^a + \frac{1}{2} \sum_i < R(f_i), e_i >
\end{aligned}
$$

The last term is half the trace of R on $\Omega^{0,1}$ or half the trace of R on the holomorphic tangent-bundle $T_X^{1,0}$. The first term can be rewritten in the old notation, where now f_i, e_i denote dual basis of $T_X \cong \Omega_X$, as

$$
\frac{1}{4} \sum_{k,l} < R(e_k), e_l > f_k f_l
$$

Thus

$$
\begin{aligned}
D^2 &= \Delta + \sum_{i<j} f_i\, f_j \otimes \left(R_{ij}(E) + \frac{1}{2} \operatorname{tr}\left(R_{ij} | T_X^{1,0} \right) \right) \\
&\quad + \frac{1}{8} \sum_{i,j,k,l} < R_{ij}(e_k), e_l > f_i f_j f_k f_l
\end{aligned}
$$

In the last term the curvature R_{ij} acts on Ω_X. If we instead consider it on T_X this term changes sign. Using the identity

$$R_{i,j}(e_k) + R_{j,k}(e_i) + R_{k,i}(e_j) = 0$$

(sketch of a proof: Assume that $e_i = \partial/\partial x_i$ are coordinate-derivatives. Then the left hand side is the sum of $\nabla_i(\nabla_j\, e_k - \nabla_k\, e_j)$ and its cyclic permutations)
one easily sees that this becomes a scalar, equal to

$$-\frac{1}{4}\sum_{i,j} < R_{ij}(f_i), f_j >$$

This completes the proof of the theorem.

Choose local coordinates x_i in which the metric g_X is given by the unit-matrix, up to order ≥ 2, and let $e_i = \partial/\partial x_i$. Then in the basis given by e_i the leading terms in $\nabla_i = \partial_i + \Gamma_i$ is given by

$$\Gamma_i = -\frac{1}{2}\sum_j R_{ij} \cdot x_j + \text{order} \geq 2$$

where $R_{ij} = \sum_{k,l} R_{ijk,l}\, e_k e_l + \text{order } 0$.

Thus the term of Clifford-filtration three in D is given by

$$-\frac{1}{2}\sum_{i,j,k,l} R_{ijk,l}\, e_i x_j e_k e_l$$

By the symmetries of the Riemann-tensor this has Clifford-degree ≤ 1. This also illustrates the remarkable cancellation which follow from the Kähler condition.

INDEX THEOREM -ABSOLUTE CASE

Let X be a compact Kähler manifold of dimension n. Let E be a hermitian vector bundle on X. We see from theorem 4.2 that the Dirac operator $D^2 = D^2$ is an elliptic and self adjoint operator on $S(X) \otimes E$. We write $\exp(-tD^2)(x, x)$ for the values at the diagonal of its heat kernel. This is an element in $\text{End}(S(X) \otimes E)$. It is of interest because of the following global index theorem:

Theorem 4.3:

$$\mathrm{tr}_s\, e^{-D^2} = \int_X \mathrm{tr}_s\, e^{-tD^2}(x,x)dx = \chi(E)$$

where $\chi(E)$ is the Euler characteristic of E:

$$\chi(E) = \sum (-1)^i H^i(X,E).$$

Proof: Consider the operator $P = 1 + D^2$ on $L^2(X, S \otimes E)$. By Garding's inequality we see that P^{-1} is a compact operator on L^2. Since D^2 is self-adjoint this implies that L^2 has a decomposition into eigenspaces of P^{-1} and also of D^2. Decomposing L^2 into eigenspaces we see that all contributions cancel except those from the kernel of D^2, which give the Euler-Poincaré characteristic $\chi(E)$.

Although there is no simple formula for $\mathrm{tr}_s\, e^{-tD^2}(x,x)$, we can at least compute its limit as $t \to 0$.

For this consider $e^{-tD^2}(x,x)$ as an element of $C(X) \otimes \mathrm{End}(E) \otimes \Omega_X^{n,n}$.

Theorem 4.4: $\lim_{t\to 0} \mathrm{tr}_s\left(e^{-tD^2}(x,x)\right) =$ *degree (n,n) - part of* $\mathrm{Td}'(T_X^{1,0}) \cdot \mathrm{ch}'(E)$.

Proof: We use the scaling and perturbation-expansion of the previous chapter. Choose local coordinates $z_j = x_j + i \cdot x_{j+n}$ near x such that the $\partial/\partial z_j$ form an ON-basis of $T_X^{1,0}$, up to order ≥ 2. Then the $\partial/\partial x_j$ have norm $\sqrt{2}$, so the corresponding Laplacian on the tangent-space is $\frac{1}{2}\sum_j (\partial/\partial x_j)^2$. Furthermore choose a local ON-basis for $\Omega_X^{0,1}$ (and thus $S(X)$) and E near x, parallel up to first order. The ON-basis of $\Omega_X^{0,1}$ will in general be different from the $d\bar{z}_j$. In these basis the connections have the form

$$\nabla_j = \partial/\partial x_j + \Gamma_j,$$

where $\Gamma_j \in C(X) \otimes \mathrm{End}(E)$ has Clifford-degree ≤ 2 and vanishes at the origin. One derives that $\Gamma_j = \frac{1}{2}\sum_k R_{jk} \cdot x_k +$ higher order.

In the perturbation expansion we first have to scale coordinates by $t^{\frac{1}{2}}$. Thus we have new coordinates $x'_j = t^{-\frac{1}{2}} \cdot x_j$, and derivatives $\partial'_j = t^{\frac{1}{2}} \cdot \partial_j$. Also in the x'_j the curvature gets a factor t, which however is cancelled if we represent it by an element of the Clifford-algebra acting on $E \otimes S(X)$. We derive from the Lichnerowicz-formula that in these new coordinates $t D^2$ has the form

$$-\frac{1}{2}\sum_j \left(\frac{\partial}{\partial x_j} + \frac{t}{2}\sum_k R_{jk}(X)x'_k\right)^2 + t\sum_{i<j} dx_i dx_j \left(R_{ij}(E) + \frac{1}{2}\mathrm{tr}\, R_{ij}(T_X^{1,0})\right)$$
$$+ \text{ higher order}$$

The "higher order" terms have a t-power which is greater than one half of their Clifford-degree. Here the R_{ij}'s are considered as elements of $\mathrm{End}(E \otimes S(X))$. If we represent them by elements of $C(X) \otimes \mathrm{End}(E)$ they lose their factor t.

Also by scaling the super-commutators in $C(X)$ we obtain a factor t, so that in the limit $t \to 0$ we can work in the exterior algebra $\wedge^* \Omega_X$.

Now in the perturbation-series we obtain elements of Clifford- degree $2n$ only with a power t^n in front, and no higher t-power of the "higher order" terms contributes. Thus for the limit we may neglect these and consider the operator (defined as a perturbation-series)

$$\exp\Big\{ \frac{1}{2} \sum_j \Big(\frac{\partial}{\partial x_j} + \frac{1}{2} \sum_k R_{jk} \cdot x_k\Big)^2 + \sum_{i<j} dx_i dx_j \big(R_{ij}(E) + \frac{1}{2} \operatorname{tr} R_{ij}(T_X^{1,0})\big) \Big\}$$

Here the R_{jk} and dx_i, dx_j are considered as elements of the Grassmann-algebra $\wedge^* \Omega_X$. The second summand in the exponent contributes up to some scaling a factor $ch'(E) \cdot e^{\frac{1}{2} c_1(T_X^{1,0})}$. For the first one we proceed as follows, following Getzler:

For indeterminates R_{jk} satisfying $R_{jk} = -R_{kj}$ consider the value at $(0,0)$ of the kernel of

$$(2\pi)^n \cdot \exp\Big(\frac{1}{2} \sum_j \Big(\frac{\partial}{\partial x_j} + \frac{1}{2} \sum_k R_{jk}\, x_k\Big)^2\Big).$$

This is a power-series $\Phi_n(R_{jk})$, with constant term 1, whose coefficients are polynomials in the R_{jk}. Furthermore this series is invariant under the orthogonal group $0(2n, \mathbb{R})$, acting on the R_{jk} by identifying them with the entries of a matrix in the Lie-algebra of $so(2n)$.

Such a matrix has eigenvalues $\pm\lambda_1, \ldots, \pm\lambda_n$, and the invariant polynomials are polynomials in the symmetric functions of the $\lambda_i{}^2$. Furthermore by decomposing $\mathbb{R}^{2n}$ into an orthogonal direct sum of even-dimensional subspaces we see that there is an even Power-series $\Phi(t)\epsilon 1 + t^2 \cdot \mathbb{R}[[t^2]]$ such that

$$\Phi_n(tR_{jk}) = \prod_{i=1}^n \Phi(t\,\lambda_i)$$

Now in our case the R_{jk} define the curvature-form of T_X and $T_X^{1,0}$ (using $SO(2n\,\mathbb{R}) \supseteq U(n, \mathbb{C})$), if we take into account the known symmetries of the curvature-tensor. Finally to compute the super-trace tr_s we have to collect terms of Grassmann-degree $2n$. Up to some scaling-factor these are given as follows: The power-series Φ defines a characteristic class $\Phi(T_X^{1,0})$ of the holomorphic tangent-bundle of X. Then the super-trace in question is equal to the highest-degree term in

$$\Phi(T_X^{1,0}) \cdot \mathrm{ch}'(E) \cdot e^{\frac{1}{2}c_1(T_X^{1,0})}$$

Thus $\chi(E)$ is obtained by integrating this over X. If we apply this to projective algebraic manifolds X and compare to the Riemann-Roch we obtain that there are no scaling-factor, that $\Phi(t) = \frac{t/2}{\sinh(t/2)}$ defines the $\hat{A}$-genus, and finally that the assertion of theorem 4.4 holds.

Remark 4.5.

  a) One shows in the same way the following generalization: Suppose α is a (p,p)-form on X, defining an element α^a in $C(X)$. Then

$$\lim_{t \to 0} \mathrm{tr}_s \left((2\pi i t)^p \cdot \alpha^a e^{-tD^2} \right) = \int_X \alpha \cdot \mathrm{Td}(T_X^{1,0}) \cdot \mathrm{ch}(E)$$

  The factor $2\pi i t$ comes from the fact that in one complex variable the heat-kernel has as leading term $(4\pi i t)^{-1} d\bar{z} \wedge dz$, and that $d\bar{z} \wedge dz$ has supertrace 2.

  b) We have also computed the series Φ. We can apply this result to any Laplacian of the form

$$\frac{1}{2} \sum_j \left(\frac{\partial}{\partial x_j} - \frac{1}{2} \sum_k R_{jk} \cdot x_k \right)^2.$$

SUPER CONNECTIONS

Suppose $E = E_+ \oplus E_-$ is a superbundle (i.e. $\mathbb{Z}/2\mathbb{Z}$-graded) on a C^∞-manifold X. A superconnection on E is an odd X-endomorphism ∇ of $E \otimes \wedge^* \Omega_X$, satisfying the usual connection rule. Then $\nabla^2 \in \mathrm{End}(E) \otimes \wedge^* \Omega_X$ is $\wedge^* \Omega_X$-linear and even. Locally a superconnection is the sum of an ordinary connection and an odd $\wedge^* \Omega_X$-linear endomorphism.

A holomorphic analogue is given as follows: Suppose X is a complex manifold. $\mathbb{Z}$-grade $\wedge^* \Omega_X$ by the rule that $\Omega^{p,q}$ has degree $q - p$. If $E = \oplus_{n \in \mathbb{Z}} E^n$ is a $\mathbb{Z}$-graded C^∞-bundle, then ∇ should be a superconnection which is a sum $\nabla = \nabla' + \nabla''$, where ∇' has degree - 1 and ∇'' degree + 1. For example if $(E^\cdot, v)$ is a complex of holomorphic hermitian bundles,

∇_0 the hermitian connection on $E = \oplus E^n$, then $\nabla = \nabla_0 + v + v^*$ is a super connection, $\nabla' = \nabla'_0 + v^*, \nabla'' = \nabla''_0 + v$. If v is holomorphic then $\nabla'^2 = \nabla''^2 = 0$, so that ∇^2 has degree 0. We say in general that ∇ is holomorphic if ∇^2 preserves degrees.

Suppose $f: X \longrightarrow Y$ is a proper and smooth holomorphic map of Kähler-manifolds of dimension n respectively r, and E a holomorphic hermitian bundle on X. We intend to define a holomorphic superconnection $\widetilde{\nabla}$ on the C^∞- direct image $f_*(E \otimes \wedge^* \Omega^{0,1}_{X/Y})$ or better $f_*(E \otimes S(X/Y))$, considered as an infinite-dimensional bundle on Y. This superconnection will be a limit of the holomorphic connections associated to the metrics $g_X + s^{-1} g_Y$ on T_X, as $s \to 0$. There are two ways to describe $\widetilde{\nabla}$:

a) Let $S(Y) = \wedge^* \Omega^{0,1}_Y$ and $S(X) = \wedge^* \Omega^{0,1}_X$ denote the spin-bundles, equipped with the metrics scaled by s (i.e. on $\Omega^{0,1}_Y$ the metric has a factor s, etc.). Let D_s denote the Dirac-Operator associated to $f^*(S(Y))^* \otimes E$, that is D_s is an endomorphism of $\mathcal{H}om(f^* S(Y), E \otimes S(X))$. (Although this is not really necessary, one can endow the dual of $S(Y)$ with the holomorphic structure given by identifying it (via the metric) with its complex conjugate $\wedge^* \Omega^{1,0}_Y$.) Now Ω_X splits orthogonally into $\Omega_X \cong f^* \Omega_Y \oplus \Omega_{X/Y}$, where the first direct summand is a holomorphic subbundle. Also the splitting is independent of s and g_Y. This induces $S(X) \cong f^*(S(Y)) \otimes S(X/Y)$, and

$$\mathcal{H}om\left(f^*(S(Y)), E \otimes S(X)\right) \cong E \otimes C(Y) \otimes S(X/Y)$$

As $s \to 0$ the Clifford-algebra $C(Y)$ converges to the exterior algebra $\wedge^* \Omega_Y$, and the Dirac-operators D_s have a limit which is the superconnection $\widetilde{\nabla}$ acting on the C^∞-sheaf $E \otimes f^* \Omega^*_Y \otimes S(X/Y)$ and also its direct image under f_*.

From this description we derive that $\widetilde{\nabla}$ is holomorphic (as D_s has degrees ± 1 and D_s^2 degree 0). Also on $f_*(E \otimes \wedge^* \bar{\Omega}_X)$ the antiholomorphic part $\widetilde{\nabla}''$ coincides with $\bar{\partial}$, which determines it uniquely and also shows that $\widetilde{\nabla}''$ commutes with base change $Y_1 \to Y$ (replacing X by $X_1 = X \times_Y Y_1$), and is independent of g_Y. Applying this to $E^* \simeq \bar{E}$ and using complex conjugation we derive that this remains true for $\widetilde{\nabla}'$. (The complex conjugate of $E \otimes S(X)$ is C^∞-isomorphic to $E^* \otimes \Omega^{n,0}_X \otimes S(X)$, respecting connections.)

However we still have to check that the limit actually exists. To do this we use the other description.

b) Let $x \epsilon X$ and $y = f(x) \epsilon Y$. Choose local coordinates $z - \alpha$ near y such that dz_α is parallel on Y at y, i.e. g_Y is given by a Kähler-potential $\varphi_Y = \sum_\alpha |z_\alpha|^2 + (\text{order} \geq 4)$. Then extend the z_α (or better $z_\alpha \circ f$) by local coordinates w_i near x such that the dw_i form an ON-basis of $\Omega_{X/Y}$ at X, and are orthogonal $f^*\Omega_Y$. Finally we modify the w_i in such a way that the Kähler-potential φ_Y is of the form $\varphi_Y = \sum_i |w_i|^2 + (\text{terms of order} \geq 2 \text{ in } z_\alpha) + (\text{order} \geq 4)$.

If we use the $d\bar{w}_i$ and $d\bar{z}_\alpha$ as a local basis for $S(X) = \wedge^*\Omega_X^{0,1}$, then in this basis the covariant derivatives are of the form $\nabla_\mu = \partial_\mu + \Gamma_\mu$ ($\mu = \alpha$ or i) where ∂_μ denotes the usual derivative and the Christoffel-symbol $\Gamma_\mu \epsilon F^2(C(\Omega_X))$ is given by third-order derivatives of φ_X. If we scale by s we obtain $\varphi_{X,s} = \varphi_X + s^{-1} f^* \varphi_Y$. So $\Gamma_{\mu,s} = \Gamma_{\mu,X} + s^{-1} f^* \Gamma_{\mu,Y}$. However the action of $\Gamma_{\mu,Y}$ on $C(Y)$ is via commutators, which introduces an extra factor s and thus the limit exists.

This description is especially simple if Y is flat near y (which we can always achieve by adding a term $\partial\bar{\partial}\varphi$ to φ_Y): Write $\nabla_\mu = \partial_\mu + \Gamma_\mu$ with $\Gamma_\mu \epsilon C(X)$, and let $s \to 0$: Then $C(X)$ converges to $C(X/Y) \otimes \wedge^*\Omega_Y$, and the Γ_μ to $\tilde{\Gamma}_\mu$. Thus the superconnection is given by $\tilde{\nabla}_\mu d = \partial_\mu + \tilde{\Gamma}_\mu$. However one has to be careful: In general (if Y is not locally flat) there is no well-defined $\tilde{\nabla}_\mu$. Only the combination $\sum_\mu e_\mu \tilde{\nabla}_\mu$ exists.

Thus:

Theorem 4.6: $\tilde{\nabla}$ *is holomorphic, independent of* g_Y, *and commutes with base-change.*

There is also a "super-Lichnerowicz formula," where the Laplacian becomes $-\sum_i \tilde{\nabla}_i^t \circ \tilde{\nabla}_i$, the summation running over on ON-frame e_i for T_Z ($Z = $ fibre of f). Note that the change of volume-forms from X to Z does not change the definition of $\tilde{\nabla}_i^t$, as the e_i respect the volume on Y.

Also note that the $\tilde{\nabla}_i$ changes Grassmann-degrees by at most two, and the degree two part is determined by the difference between the pullback-metric on f^*T_Y and the quotient-metric (from T_X). Finally the part of degree one codifies the fact that the extention $0 \to T_Z \to T_X \to f^*T_Y \to 0$ does not split metrically.

Finally we can relate the heat-kernels for D_s^2 and $\tilde{\nabla}^2$. Note that as operator on $f_*(E \otimes S(X/Y)) \otimes \wedge^*\Omega_Y$ $\tilde{\nabla}^2$ is of the form $\nabla_{\bar{\partial}_{X/Y}} + $ nilpotent terms. Thus by perturbation-theory we may define the supertrace $\text{tr}_s(\exp(-\tilde{\nabla}^2))$, which is a C^∞-section of $\wedge^*\Omega_Y$.

We define $\text{ch}'(f_*(E \otimes S(X/Y)))$ as the C^∞-section of $\wedge^*\Omega_Y$ which in degree (p,p) differs from $\text{tr}_s e^{-\tilde{\nabla}^2}$ by a factor $(\frac{1}{2\pi i})^p$. We then have:

Lemma 4.6:

$$\lim_{s \to 0} \int_{X/Y} \text{tr}_s \, e^{-D_s^2}(x, x)dx = \text{degree}(r, r) - \text{part of } ch'(f_*(E \otimes S(X/Y))) \cdot \text{Td}_Y \, .$$

Proof: We now use the relative perturbation argument of the previous chapter: Choose local coordinates z_i near $y \in Y$ such that the $\partial/\partial z_i$ form on a parallel ON-basis of T_Y up to order ≥ 2, and also choose local frames. Finally in the Lichnerowicz-formula for D_s^2 scale the z_i-coordinates by $\sqrt{s}$, and compare the powers of s and the Clifford-degrees relative Ω_Y, i.e. the number of factors Ω_Y in $C(X) = C(X) = C(Y) \otimes C(X/Y)$: One finds again that the s-power is at least half the Clifford-degree. By the same argument as before we then may replace the Clifford-algebra $C(Y)$ by the exterior algebra $\wedge^* \Omega_Y$, and do perturbation theory. The relevant operator is then the sum of $\tilde{\nabla}^2$, the horizontal Laplacian, and the horizontal contribution to $c_1(T_X^{1,0})$.

To describe the horizontal Laplacian we use local coordinates z_α as before (so that $\varphi_Y = \sum_\alpha |z_\alpha|^2 + (\text{order} \geq 4)$), and extend by adding w_i's. One then sees that up to terms disappearing in the scaling limit we may replace the Christoffel-symbols Γ_α on X by those on Y.

Now our perturbation-calculation proceeds as before, except that now on Y we have the infinite-dimensional bundle $f_*(E \otimes S(X/Y))$, whose curvature is given formally by $\tilde{\nabla}^2$. However all the necessary algebraic identities continue to hold in this context, and so we obtain the same result. Namely the super trace converges to

$$ch'(f_*(E) \otimes S(X/Y)) \cdot \text{Td}_Y \, .$$

So the lemma has been shown.

Remark 4.8.

Similar we obtain the following generalization: Assume that α is a (p, p)-form on Y. Then

$$\lim_{s \to 0} (2\pi i s)^p \int_{X/Y} \text{tr}_s \left(\alpha^a e^{D_s^2}(x, x)\right)dx = \text{highest-degree part in}$$

$$ch'\left(f_*(E \otimes S(X/Y))\right) \cdot \alpha \cdot \text{Td}_Y \, .$$

The relative index theorem.

We now also scale the metric on X, replacing g_X by $t^{-1}g_X$ and letting $t \to 0$. This way we obtain a family of superconnections $\widetilde{\nabla}_t$ on $f_*(E \otimes S(X/Y))$ and chern-characters $\mathrm{ch}'(f_*(E \otimes S(X/Y)))$. By the previous chapter these have asymptotic expansions in t. Note that $\widetilde{\nabla}_t^2$ differs from $t \cdot \widetilde{\nabla}_1^2$ by a factor $t^{-\frac{1}{2}}$ for each Grassmann-degree in Ω_Y.

Theorem 4.9:

$$\lim_{t \to 0} \mathrm{ch}'\left(f_*(E \otimes S(X/Y))\right) = \int\limits_{X/Y} \mathrm{ch}'(E) \cdot \mathrm{Td}(T_X^{1,0})$$

Proof: It suffices to show that both sides have the same integrals over Y against any (p,p)-form α on Y, with compact support. Instead we also may integrate against $\alpha \cdot \mathrm{Td}_Y$. By the previous lemma on the left we obtain for a fixed $t > 0$ the limit

$$\lim_{s \to 0}(2\pi st)^p \int\limits_X \mathrm{tr}_s(\alpha^a e^{-tD_s^2})(x,x)dx,$$

(Taking into account that the action of α^a is scaled by t^p).

By remark 4.5 the inner term converges for fixed s and $t \to 0$ to $\int_X \alpha\, \mathrm{ch}'(E)\, \mathrm{Td}_X$. (Again the α^a-action is scaled as it depends on the metric. Now the factor is s^p.) On the other hand one checks that as $s \to 0$ Td_X converges to $f_*(\mathrm{Td}_Y) \cdot \mathrm{Td}_{X/Y}$. Thus on the left we obtain

$$\int\limits_X \alpha\, \mathrm{ch}(E)\, \mathrm{Td}_{X/Y} \cdot f^*(\mathrm{Td}_Y),$$

which coincides with the integral on the right.

COMPUTATION OF SUBDOMINANT TERMS

We have seen that $\mathrm{tr}_s(\alpha^a \cdot e^{-tD^2})$ has an asymptotic expansion in t. The leading term has been computed. Here we want to consider the next term. However we can do this only if α is a closed (p,p)-form, i.e., $d\alpha = 0$. To describe the result we first need some notation:

Denote by ω_X the Kähler-form on X. If dz_j is a local ON-base of $T_X^{1,0}$, then

$$\omega_X = \frac{i}{2\pi} \sum_j dz_j \wedge d\bar{z}_j$$

If $\delta\omega_X$ is a real $(1,1)$-form we can change the metric g_X to $g_X + \epsilon\delta g_X$ (ϵ small) such that $\omega_X\hat{A}$ changes to $\omega_X + \delta\omega_X$. In this procedure the $\hat{A}$-class $\hat{A}_X$ changes by $\epsilon \cdot \delta\hat{A}_X$, and the theory of lecture 2 provides us the a class $\hat{A}(\delta\omega_X)$, uniquely defined modulo $\operatorname{Im}(\partial) + \operatorname{Im}(\bar{\partial})$, such that up to higher order in ϵ

$$\delta\hat{A}_X = \frac{\partial\bar{\partial}}{\pi i}\hat{A}_X(\delta\omega_X)$$

We extend $\widetilde{\hat{A}}_X(\delta\omega_X)$ to arbitrary $(1,1)$-forms by linearity. Finally if α is a (p,p)-form, we define as follows $\widetilde{\hat{A}}_X(\alpha)$:
Express α in a local ON-basis dz_j, and write for any pair (j,k) $\alpha = dz_j \wedge d\bar{z}_k \wedge \alpha_{jk}+$ terms not involving dz_j or not involving $d\bar{z}_k$. Then

$$\widetilde{\hat{A}}(\alpha) = \sum_{j,k} \widetilde{\hat{A}}_X(dz_j \wedge d\bar{z}_k) \wedge \alpha_{jk}$$

With this notation we have

Theorem 4.10: *Assume α is a closed (p,p)-form. Then*

$$2\pi i \cdot \lim_{t\to 0} \frac{\partial}{\partial t} (2\pi it)^p \operatorname{tr}_s(\alpha^a e^{-tD^2}) = 2 \cdot \int_X \widetilde{\hat{A}}(\alpha) \cdot e^{\frac{1}{2}c_1(T_X)} \operatorname{ch}'(E)$$

Proof: We show that this holds up to an universal factor. That this factor must be 1 will be seen later, be computing an example. We compute as follows, using the fact the tr_s vanishes on super commutators. In the following da, db will denote Grassmann-variables, and $(\ )_{db}$ will denote the term involving db, etc.:

$$\frac{\partial}{\partial t}\,\mathrm{tr}_s(t^p\alpha^a e^{-tD^2}) = \frac{\partial}{\partial t}\,\frac{\partial}{\partial b}\Big|_{b=0}\,\mathrm{tr}_s(e^{-tD^2+bt^p\alpha^a})$$

$$= \frac{\partial}{\partial b}\Big|_{b=0}\,\mathrm{tr}_s\big((-\tfrac{1}{2}\{D,D\}+pbt^{p-1}\alpha^a)e^{-tD^2+bt^p\alpha^a}\big)$$

$$= \mathrm{tr}_s(pt^{p-1}\alpha^a e^{-tD^2}) + \frac{\partial}{\partial b}\Big|_{b=0}\,\mathrm{tr}_s(-\tfrac{1}{2}D\{D,e^{-tD^2+bt^p\alpha^a}\})$$

$$= \mathrm{tr}_s(e^{-tD^2+pt^{p-1}\alpha\,da\,db})_{da\,db} + \mathrm{tr}_s(\tfrac{1}{2}De^{-tD^2+t^p\{D,\alpha^a\}db})_{db}$$

$$= \mathrm{tr}_s(e^{-tD^2-\frac{1}{2}D\,da+t^p\{D,\alpha^a\}db+pt^{p-1}\alpha^a\,da\,db})_{da\,db}$$

$$= \mathrm{tr}_s(e^{-tD^2-\frac{\sqrt{t}}{2}D\,da+t^{p-\frac{1}{2}}\{D,\alpha^a\}db+pt^{p-1}\alpha^a\,da\,db})_{da\,db}$$

The exponent can be written as in a Lichnerowicz-formula. We define a super-connection ∇'_j with coefficients in $\mathbb{C}[da,db]$ by the following rule: Choose a local ON-basis e_j of T_X. Then

$$\nabla'_j = \nabla_j - \frac{t^{-\frac{1}{2}}}{2}e_j\,da + \frac{t^{p-\frac{1}{2}}}{2}\{e_j,\alpha^a\}db$$

If we assume in addition that $\nabla_j(e_k) = 0$ at the origin, then we compute, using that $\sum\limits_j e_j\,\nabla_j(\alpha^a)$ has degree $\leq 2p-1$ (as $d\alpha = 0$) and that

$$\sum_j (e_j\{e_j,\alpha^a\} - \{e_j,\alpha^a\}e_j) = -4p\alpha^a$$

(expand α as monimial in e's).
that

$$t\sum_j \nabla'^2_j = t\sum_j \nabla^2_j - \frac{t^{\frac{1}{2}}}{2}D\,da + t^{p-\frac{1}{2}}\{D,\alpha^a\}db + p\,t^{p-1}\alpha^a\,da\,db$$
$$+ (?)\,db,$$

where (?) is a term which has Clifford-degree $\leq 2p-1$, has a $t^{p-\frac{1}{2}}$ in front, and involves no da. It thus follows that the exponent in our previous formula is of the type Laplacian + curvature, where the Laplacian is formed with ∇'_i instead of ∇. Thus the limit as $t \to 0$ should be given by a product $\hat{A}$-class $\cdot$ Chern-class. However, we first have to clear up a small point: If in local coordinates $\nabla'_j = \partial_j + \Gamma_j$, the Christoffel-symbols Γ_j do not vanish at the origin. However, if we formally conjugate with $\exp(\sum\limits_j x_j\,\Gamma_j(0))$ we

do not change the trace, remove the constant term, and do not disturb the curvature-term because commuting with $\Gamma_j(0)$ reduces Clifford-degrees (as the entries are of the form (degree one) $\wedge\, da$ or (degree one) $\wedge\, db$). Also commuting with a quadratic exponent we get after conjugation

$$\nabla'_j = \partial_j + \frac{1}{2}\sum_k x_j \cdot R'_{jk} + \text{ higher order,}$$

where $R'_{jk} = [\nabla'_j, \nabla'_k]$ denotes the curvature. In our case we compute

$$R'_{jk} = R_{jk} - \frac{t^{p-1}}{4}(\{e_j, \{e_k, \alpha^a\}\} - \{e_k, \{e_j, \alpha^a\}\})\, da\, db$$
$$+ (??)\, db,$$

where the $(??)\, db$-term will disappear at the end. Namely the limit as $t \to 0$ is obtained by substituting R'_{jk} into the Todd-class (considering R'_{jk} as a show-symmetric matrix) and the curvature of E + the term $(?)\, db$ into the chern-class. The db-terms disappear because they find no da, so finally we only have to find the $dadb$-term in

$$\int_X \hat{A}(R'_{jk}) \cdot e^{\frac{1}{2}\, c_1(T_X)} \cdot \text{ch}(E).$$

It remains to use linear algebra to see that this is what we want it to be.

First of all one easily sees that it suffices to check the assertion for (1,1)-forms, as the extension to higher degrees is the same on both sides. Now if α is a (1,1)-form, we calculate in a local good basis z_j: α determines a change in metric δg_X, and the change in curvature corresponding to it is given by $\hat{A}$ the derivatives $\partial\bar{\partial}\delta g_X$. Thus the change $\delta \hat{A}'_X$ is equal to a derivative of the $\hat{A}$-polynomial, multiplied with $\partial\bar{\partial}\delta g_X$. One derives that $\tilde{\hat{A}}(\alpha)$ is proportional to the derivative at $\epsilon = 0$ of $\hat{A}'(R_X + i\epsilon\, Q)$, where R_X is the curvature of X and δg_X is given by a hermitian endomorphism Q of $T_X^{1,0}$. But this just corresponds to the change from R_{jk} to R'_{jk}.

Thus finally the theorem has been shown, up to a scalar factor which we shall work out later.

The Relative Case

Finally we show a relative version of Theorem 4.9. Let $f: X \to Y$ denote a smooth map of compact Kähler- manifolds, and consider the metric

$g_{X,s} = g_X + s^{-1} f^* g_Y$ on X. If α is a (p,p)-form on X the action α^a depends on s, so we denote it by α_s^a.

We also obtain a superaction $\widetilde{\alpha}^a$ on $f_*(E \otimes S(X/Y))$. According to the decomposition $\Omega_X = f^* \Omega_Y \oplus \Omega_{X/Y}$ elements of $\wedge^* \Omega_Y$ act by Grassmann-multiplication, and $\wedge^* \Omega_{X/Y}$ by Clifford-action as before. If we scale g_X to $t^{-1} g_X$, the action of $f^* \Omega_Y$ gets a factor $t^{-\frac{1}{2}}$.

As before the supertrace $\mathrm{tr}_s(\widetilde{\alpha}^a \cdot e^{-\widetilde{\nabla}^2})$ of $\widetilde{\alpha}^a \cdot e^{-\widetilde{\nabla}^2}$ on $f_*(E \otimes S(X/Y))$ can be obtained by passing to the limit $s \to 0$, if we scale it by factor $(\frac{1}{2\pi i})^p$ in degrees (p,p):

$$\lim_{s \to 0} \int_{X/Y} \mathrm{tr}_s(\alpha_s^a \cdot e^{-D_s^2})(x,x) = \mathrm{tr}_s(\widetilde{\alpha}^a \cdot e^{-\widetilde{\nabla}^2}) \cdot \mathrm{Td}_Y',$$

up to scaling.

Now we also change the metric g_X to $t^{-1} g_X$, and let $t \to 0$. We obtain:

Theorem 4.11:

Suppose α is a closed (p,p)-form on X. Let $\mathrm{tr}_s(\widetilde{\alpha}^a \cdot e^{-\widetilde{\nabla}_t^2})^{\mathrm{scaled}}$ denote the supertrace modified by multiplying the (p,p)-part by $(\frac{1}{2\pi i})^p$. Then

$$2\pi i \lim_{t \to 0} \frac{\partial}{\partial t} (2\pi i t)^p \, \mathrm{tr}_s(\widetilde{\alpha}^a \cdot e^{-\widetilde{\nabla}_t^2}) = 2 \int_{X/Y} \widetilde{\hat{A}}_{X/Y}'(\alpha) \cdot e^{\frac{1}{2} c_1(T_{X/Y})} \cdot \mathrm{ch}'(E).$$

modulo $\mathrm{Im}(\partial) + \mathrm{Im}(\bar{\partial})$

Proof: It suffices to show that both sides have the same integral against any closed (q,q)-form β on Y (using that Y is Kähler). But then the assertion amounts to theorem 4.10, applied to $\alpha \wedge f^*(\beta)$.

LECTURE 5. NUMBER OPERATORS
AND DIRECT IMAGES

In this lecture we define the direct image, for a smooth proper map of arithmetic varieties. The idea is as follows: If E is an acyclic bundle on X, endow $F = f_*(E)$ with some hermitian metric. Then $f_*([E])$ is the sum of $[F]$ and a secondary characteristic class which measures the difference between F and $f_*(E \otimes S(X/Y))$. To define this class we need number-operators and various estimates for Laplacians, which make up the bulk of this chapter. There is also such a definition in [GS3]. I do not know whether it coincides with ours. At the end we consider composition of maps. References: [BFS1], [GS3], [Q1]

NUMBER-OPERATORS

We start with the finite-dimensional case. Suppose X is a complex manifold and $(E^\bullet, v)$ a finite complex of holomorphic vectorbundles on X ($v: E^n \to E^{n+1}$ denotes the differential, which is a holomorphic map). Suppose each E^n has a hermitian metric, and thus $E = \oplus E^n$ has a metric and and connection ∇. Extend v and its adjoint v^* to $E \otimes \Omega_X$ by the rule that they supercommute with Ω_X. Then $\nabla_v = \nabla + v + v^*$ becomes a superconnection on E, which is holomorphic as ∇_v^2 is of type $(1,1)$ (the part of hype $(0,2)$ is $\nabla'' v + v \nabla'' = 0$, and similar for type $(2,0)$). One thus derives that $\mathrm{tr}_s(\exp(-\nabla_v^2))$ represent the Chern-character $\mathrm{ch}(E^\bullet) = \sum_n (-1)^n \mathrm{ch}(E^n)$ in cohomology. We want to study its behavior under change of metrics. However we only treat infinitesimal variations:

A one-parameter family of metrics on $E^\bullet$ is of the form $< e_1, e_2 >_\epsilon = < e_1, e_2 > + \epsilon \cdot$
$< e_1, N(e_2) > + O(\epsilon^2)$, where N is a selfadjoint endomorphism of $E^\bullet$ respecting the grading. For the connection $\nabla_v'' = \nabla'' + v$ remains unchanged, while $\nabla_v' = \nabla' + v^*$ changes by $-\epsilon\{N, \nabla_v\}$. We get a more symmetric picture by conjugating with $1 + \frac{\epsilon}{2} N$, which does not change traces. Then the variation in connections is given by

$$\delta \nabla_v = \frac{\epsilon}{2}\{N, \nabla_v'' - \nabla_v'\}$$

Thus we compute:

$$\frac{\partial}{\partial \epsilon}\Big|_{\epsilon=0} \text{tr}_s \left(\exp(-\nabla_{v,\epsilon}^2) \right) = \frac{\partial}{\partial \epsilon}\Big|_{\epsilon=0} \text{tr}_s \left(\exp(-\nabla_v^2 - \frac{\epsilon}{2}\{\nabla_v, \{N, \nabla_v'' - \nabla_v'\}\}) \right)$$

$$= -\frac{1}{2}\text{tr}_s \left(\{\nabla_v\{N, \nabla_v'' - \nabla_v\}\} \exp(-\nabla_v^2) \right)$$

$$= \frac{1}{2}\text{tr}_s \left(\{\nabla_v, \{\nabla_v'' - \nabla_v', N\} \exp(-\nabla_v^2)\} \right)$$

$$= \frac{1}{2} d \left(\text{tr}_s \left(\{\nabla_v'' - \nabla_v', N\} \exp(-\nabla_v^2)\} \right) \right)$$

To go on we note that

$$d\left(\text{tr}_s \left(N \exp(-\nabla_v^2) \right) \right) = \text{tr}_s \left(\{\nabla_v, N\} \exp(-\nabla_v^2) \right)$$

$$= \text{tr}_s \left(\{\nabla_v', N\} \exp(-\nabla_v^2) \right) + \text{tr}_s \left(\{\nabla_v'', N\} \exp(-\nabla_v^2) \right)$$

Comparing holomorphic and antiholomorphic degrees we see that the two terms are ∂ respectively $\bar{\partial}$ applied to $\text{tr}_s \left(N \exp(-\nabla_v^2) \right)$. Thus finally

$$\frac{\partial}{\partial \epsilon}\Big|_{\epsilon=0} \text{tr}_s \left(\exp(-\nabla_{v,\epsilon}^2) \right) = \frac{1}{2} d (\bar{\partial} - \partial) \text{tr}_s \left(N \exp(\nabla_v^2) \right)$$

$$= \partial\bar{\partial} \text{tr}_s \left(N \exp(-\nabla_v^2) \right).$$

If we denote as usual $\text{ch}(E^\bullet) = \text{tr}_s \left(\exp(\frac{i}{2\pi} \nabla_v^2) \right)$ and

$$\widetilde{\text{ch}}(E^\bullet, N) = \frac{1}{2} \text{tr}_s \left(N \exp(\frac{i}{2\pi} \nabla_v^2) \right),$$

we get $\frac{\partial}{\partial \epsilon}\Big|_{\epsilon=0} \text{ch}(E_\epsilon^\bullet) = \frac{\partial\bar{\partial}}{\pi i} \widetilde{\text{ch}}(E^\bullet, N)$. That is we have found a new construction for the secondary classes of lecture 2.

We also need some compatability for a two-parameter family of metrics: Suppose we have metrics

$$< e_1, e_2 >_{\delta, \epsilon} = < e_1, e_2 > + \delta < e_1, N_1 e_2 > + \epsilon < e_1, N_2 e_2 > + \epsilon\delta < e_1, N_{12} e_2 >$$
$$+ \text{ higher order}$$

For each δ the change in ϵ is given by an infinitesimal generator

$$N_2(\delta) = \frac{\partial}{\partial \epsilon}\Big|_{\epsilon=0} (1 + \delta N_1)^{-1} (1 + \delta N_1 + \epsilon N_2 + \epsilon\delta N_{12})$$
$$= N_2 + \delta(N_{12} - N_1 N_2) \quad (+ \text{ higher order})$$

and similar $N_1(\epsilon) = N_1 + \epsilon(N_{12} - N_2 N_1)$.

Now:

Lemma 5.1.

    At $\epsilon = \delta = 0$ we have modulo $\mathrm{Im}(\partial) + \mathrm{Im}(\bar{\partial})$:

$$\frac{\partial}{\partial \delta}\ \mathrm{tr}_s\left(N_2(\delta)\exp(-\nabla^2_{v,\epsilon,\delta})\right)\ =\ \frac{\partial}{\partial \epsilon}\ \mathrm{tr}_s\left(N_1(\epsilon)\exp(-\nabla^2_{v,\epsilon,\delta})\right)$$

Proof.

    We show that the lefthand side is symmetric in 1,2. After conjugating with $1 + \frac{\delta}{2}\,N_1$ we have to compute the supertrace of

$$\frac{\partial}{\partial \delta}_{\delta=0}\left((N_2 + \delta \cdot N_{12}\ -\ \frac{\delta}{2}(N_1 N_2 + N_2 N_1))\exp(-\nabla^2_v - \frac{\delta}{2}\{\nabla_v, \{N_1, \nabla''_v - \nabla'_v\}\}\right.$$

$$= \left(N_{12} - \frac{1}{2}(N_1 N_2 + N_2 N_1)\right)\exp(-\nabla^2_v)$$

$$- \frac{1}{2}\frac{\partial}{\partial \delta}_{\delta=0}\left(N_2 \exp(-\nabla^2_v - \frac{\delta}{2}\{\nabla_v, \{N_1, \nabla''_v - \nabla'_v\}\})\right)$$

The first term is symmetric in 1,2. The second is up to a factor $-\frac{1}{2}$ equal to

$$\int_{a+b=1} N_2 \exp(-a\,\nabla^2_v)\{\nabla_v, \{N_1, \nabla''_v - \nabla'_v\}\}\exp(-b\nabla^2_v)\,da\ .$$

Up to a commutator with ∇_v this is

$$\int_{a+b=1} \{\nabla_v, N_2\}\exp(-a\,\nabla^2_v)\{\nabla'_v - \nabla''_v, N_1\}\exp(-b\nabla^2_v)da$$

The terms of type (p,p) in this are

$$\int_{a+b=1} \{\nabla''_v, N_2\}\exp(-a\,\nabla^2_v)\{\nabla'_v, N_1\}\exp(-b\,\nabla^2_v)da$$

$$- \int_{a+b=1} \{\nabla'_v, N_2\}\exp(-a\,\nabla^2_v)\{\nabla''_v, N_1\}\exp(-b\,\nabla^2_v)da$$

Their supertrace is symmetric in $(1,2)$. Thus the whole result is symmetric up to the image of d and terms of type (p,q) with $p \neq q$. As $\mathrm{tr}_s\left(N\exp(-\nabla^2_v)\right)$ is pure of type (p,p) comparing elements of this type completes the proof of the lemma.

Lemma 5.1. can be interpreted as follows: We have defined a 1-form on the space of metrics, with values in differential forms $/\operatorname{Im}(\partial) + \operatorname{Im}(\bar{\partial})$, and this 1-form is closed. Thus we obtain global secondary classes by integrating along a path connecting two metrics, and the result is independant of the path chosen.

As an example consider the case where $E^{\bullet}$ is acyclic, and thus the Laplacian $vv^* + v^*v$ has strictly positive eigenvalues.

If we multiply the metric on E^n by a factor t^n $(t > 0)$ we obtain a family of metrics with Laplacian $t(vv^* + v^*v)$, and infinitesimal generator (corresponding to td/dt) given by the number-operator N which on E^n is multiplication by n. Furthermore $\exp(-\nabla_t^2)$ decays exponentially as $t \to \infty$. It follows that formally

$$\operatorname{tr}_s(\exp\left(-\nabla_0^2\right)) = -\partial\bar{\partial}\int_0^\infty \operatorname{tr}_s\left(N\exp(-\nabla_t^2)\right)\frac{dt}{t}$$

The lefthand side is (up to $2\pi i$'s) the alternating sum of the chern-characters of the E^n. The integral on the right does not converge as it stands, as the integrand diverges as $1/t$ for $t \to 0$. However we can define it as a regularized integral, as follows:

Suppose $f(t)$ is a continuous function of $t > 0$, with sufficiently rapid decay as $t \to \infty$. Furthermore we assume that as $t \to 0$ $t^n f(t)$ is C^∞, so that t has an asymptotic expansion

$$f(t) = \sum_{l=-n}^{0} a_l t^l + O(t).$$

Then $\zeta(s) = \frac{1}{\Gamma(s)}\int_0^\infty f(t)t^{s-1}dt$ converges for $Re(s) > n$, and has an analytic extension to the whole complex plane (use partial integration). Define

$$\int_0^\infty f(t)\frac{dt}{t} = \zeta'(0)$$

For example if $f(t) = e^{-\lambda t}$, then $\zeta(s) = \lambda^{-s}$, and

$$\int_0^\infty e^{-\lambda t}\frac{dt}{t} = -\log(\lambda).$$

Also if f has a zero at the origin the integral coincides with the usual one. Finally if $f(t) = t\frac{d}{dt}g(t)$, where $t^n \cdot g(t)$ is C^∞ in 0, then

$$\int\limits_{0}^{\infty} t\frac{d}{dt}g(t)\,\frac{dt}{t} = -\lim_{t\to0}{}'g(t),$$

where on the right-hand side we have the Schwartz-limit: Let $g(t) = \sum\limits_{l=-n}^{-1} b_l t^l + \tilde{g}(t)$, then $\lim_{t\to0}{}'g(t) = \lim_{t\to0}\tilde{g}(t)$. This assertion follows by partial integration.

Thus finally we can define (for any acylic complex $E^\bullet$) $\widetilde{ch}(E^\bullet)$ as the regularized version of our class (adding $2\pi i$'s), so that formally

$$\widetilde{ch}(E^\bullet) = -\frac{1}{2}\int_0^\infty \text{tr}_s(N\exp\big(\frac{i}{2\pi}\nabla_t^2\big))\frac{dt}{t},$$

and obtain that $ch(E^\bullet) = \frac{\partial\bar\partial}{\pi i}\cdot\widetilde{ch}(E^\bullet)$.

Also for $E^\bullet$ split exact the $\widetilde{ch}$-class vanishes: We may assume that $E^\bullet$ consists of an isomorphism $E = E^n \xrightarrow{\sim} E^{n+1}$. Then $\nabla_t^2 = \nabla_0^2 + t$, and

$$\text{tr}_s\big(N\exp(\nabla_t^2)\big) = (-1)^{n+1}e^{-t}\cdot\text{tr}_s\big(\exp(-\nabla_E^2)\big).$$

As $\int\limits_0^\infty e^{-t}\frac{dt}{t} = -\log(1) = 0$, we are done.

It follows that $\widetilde{ch}(E^\bullet)$ coincides with the $\widetilde{ch}$-class of lecture II.

Finally we consider what happens if we change the metric on $E^\bullet$. For the infinitesimal version we obtain a number-operator N^*. Then (using Lemma 5.1)

$$\frac{\partial}{\partial\epsilon}_{\epsilon=0}\big(-\frac{1}{2}\int_0^\infty \text{tr}_s\big(N\exp(-\nabla_{t,\epsilon}^2)\big)\frac{dt}{t}\big)$$

$$= -\frac{1}{2}\int_0^\infty \frac{\partial}{\partial\epsilon}_{\epsilon=0}\text{tr}_s\big(N\exp(-\nabla_{t,\epsilon}^2)\big)\frac{dt}{t}\big)$$

$$= -\frac{1}{2}\int_0^\infty td/dt\,\text{tr}_s\big(N^*\exp(-\nabla_t^2)\big)\frac{dt}{t}$$

$$= \frac{1}{2}\lim_{t\to0}{}'\text{tr}_s\big(N^*\exp(-\nabla_t^2)\big),$$

modulo $\text{Im}(\partial) + \text{Im}(\bar\partial)$.

To do this computation we have to be able to integrate modulo $\text{Im}(\partial) + \text{Im}(\bar\partial)$, so this group should be closed in a reasonable topology. Thus we assume that X is compact and Kähler. However one can easily show that this is not necessary in this case.

So far we only have recovered results which could be obtained more easily by the deformation - method of lecture 2. However this method does not work for infinite- dimensional complexes like $\Gamma(X, E \otimes S(X))$, which we treat now:

Suppose we change the hermitian metric g_X, such that the Kähler-form $\omega_X = \frac{i}{2\pi} \sum dz_j \wedge d\bar{z}_j$ $(dz_j$ a local ON-basis) is changed by $\epsilon \delta \omega_X$. This effects the metric on $\Gamma(X, E \otimes S(X)$ in two ways: First the volume on X changes, and then also the hermitian metric on $\Omega_X^{0,1}$. The first effect is measured by the scalar N- operator $\operatorname{tr} Q$, where $Q \epsilon \operatorname{End}(T_X^{1,0})$ denotes the hermitian operator which infinitesimally generates the change of metrics. For the second one the N-operator is

$$-\frac{1}{2} \sum d\bar{z}_j \cdot (-Q^t dz_j) = -\frac{1}{2} \operatorname{tr}(Q) + \pi i \,\delta\omega^a$$

Thus the total N-operator is given by

$$N = \frac{1}{2} \operatorname{tr}(Q) + \pi i \cdot \delta\omega^a.$$

With this operator the previous formulas for the change of D^2 remain valid. Of course there is also a relative version:

If $f: X \rightarrow Y$ is a smooth map of compact Kähler- manifolds, we may study the metrics $g_{X,s} = g_X + s^{-1}f^*g_Y$ on X, and pass to the limit as $s \rightarrow 0$. Then the operator N as above has a limit $\tilde{N}$, where $\delta\omega_X^a$ is now an element of $C(X/Y) \otimes \wedge^*\Omega_Y$. With this $\tilde{N}$ the previous formulas describe the dependance of the limiting super-connection $\tilde{\nabla}$ on g_X. For example

$$\frac{\partial}{\partial \epsilon} \operatorname{tr}_s \left(\exp(-\tilde{\nabla}^2) \right) = \partial\bar{\partial} \operatorname{tr}_s \left(\tilde{N} \exp(-\tilde{\nabla}^2) \right)$$

or formally

$$\frac{\partial}{\partial \epsilon} \operatorname{ch} \left(f_*(E \otimes S(X/Y)) \right) = \frac{\partial\bar{\partial}}{\pi i} \operatorname{tr}_s \left(\frac{1}{2} \tilde{N} \exp(\frac{i}{2\pi} \tilde{\nabla}^2) \right)$$

For example let us compare the aymptotic expansions as we replace g_X by $t^{-1}g_X$, and let $t \rightarrow 0$:

On the left we obtain the derivative of $\int\limits_{X/Y} \operatorname{Td}(X/Y) \operatorname{ch}(E)$, which is $\frac{\partial\bar{\partial}}{\pi i}$ of

$$\int\limits_{X/Y} \widetilde{\operatorname{Td}}(X/Y, Q) \operatorname{ch}(E).$$

On the right the t^{-1} term in $\operatorname{tr}_s(\frac{1}{2}\tilde{N} \exp(\frac{i}{2\pi}\tilde{\nabla}^2))$ is equal to $\frac{1}{4} \int\limits_{X/Y} \omega_X \operatorname{Td}(X/Y) \operatorname{ch}$

which is closed.

For the t^0-term we obtain the sum of

$$\frac{1}{4} \int\limits_{X/Y} \operatorname{tr}(Q|T_{X/Y}) \ \operatorname{Td}(X/Y) \operatorname{ch}(E)$$

and

$$\int\limits_{X/Y} \widetilde{A}(T_{X/Y}, Q) \operatorname{ch}(E)$$

The first term is the contribution of the factor $e^{\frac{1}{2}c_1(T)}$ to the secondary class
$\widetilde{\operatorname{Td}}(T_{X/Y}, Q)$ associated to

$$\operatorname{Td}(T_{X/Y}) = \hat{A}(T_{X/Y}) e^{\frac{1}{2}c_1(T_{X/Y})},$$

so both terms combine to give

$$\int\limits_{X/Y} \widetilde{\operatorname{Td}}(T_{X/Y}, Q) \operatorname{ch}(E).$$

Applying $\frac{\partial \bar{\partial}}{\pi i}$ reproduces the left-hand side. This computation also proves that the factors in lecture 4 are correct: One can easily find an example where the contribution of the $\widetilde{A}$- class does not vanish.

Now suppose E is a bundle on X such that all direct images $R^i f_* E, 0 \le i < \infty$, vanish, that is E has trivial cohomology on each fibre $f^{-1}(y)$. Then formally the alternating sum of $f_*(E \otimes \Omega^{0,q}_{X/Y})$ is represented by the secondary chern-class

$$\widetilde{\operatorname{ch}}\big(f_*(E \otimes S(X/Y))\big)$$

which up to factors $2\pi i$ is given by the regularized integral

$$-\frac{1}{2} \int\limits_0^\infty \operatorname{tr}_s(\widetilde{N} \exp(-\nabla_t^2)) \, \frac{dt}{t}.$$

Here $\widetilde{N}$ denotes the number-operator associated to scaling the metric g_X to $t^{-1}g_X$, and corresponds to the infinitesimal generator $t\frac{d}{dt}$.

The integral exists as we shall see that the integrand has an asymptotic expansion for $t \to 0$, and decays exponentially as $t \to \infty$.

It also follows that

$$\frac{\partial \bar{\partial}}{\pi i} \; \widetilde{\mathrm{ch}} \; \left(f_*(E \otimes S(X/Y)) \right) = \int\limits_{X/Y} \mathrm{Td} \left(T_{X/Y} \right) \cdot \mathrm{ch}(E),$$

and that under a change of metric $\widetilde{\mathrm{ch}}$ changes by

$$\int\limits_{X/Y} \widetilde{\mathrm{Td}}(T_{X/Y}) \cdot \mathrm{ch}(E)$$

(This is true infinitesimally. Now integrate.) However we need a more general situation. Suppose now that $F = f_* E$ is locally free, and that only the higher direct images $R^i f_* F, 0 < i < \infty$, vanish. Endow F with some hermitian metric and consider the maps of complexes.

$$v \colon \Gamma(Y, F \otimes \wedge^* \Omega_Y^{0,1}) \to \Gamma(X, E \otimes \wedge^* \Omega_X^{0,1})$$

respectively

$$v \colon F \to f_*(E \otimes \wedge^* \Omega_{X/Y}^{0,1}).$$

Here we write $\wedge^* \Omega_{X/Y}^{0,1}$ instead of $S(X/Y)$ because we do not scale the metric on $\Omega_{X/Y}^{0,q}$ by $2^{q/2}$. Also for the square-integration we multiply the volume form by $(2\pi)^{\dim(Y)-\dim(X)}$. This may seem confusing. However it makes everything compatible with Serre-duality, and also simplifies the final formulas (i.e. the class $R(x)$ in lecture 6). The 2π-factor has the effect that the volume of $\mathbb{P}^1$ is normalized to 1 (and $\mathbb{P}^n$ has volume $1/n!$). On their mapping cones we consider the Dirac-operators $D_F + D_E + v + \widetilde{v}^*$ resp. $\nabla_F + \widetilde{\nabla}_E + v + \widetilde{v}^*$. Here $\widetilde{v}^*$ is the limit of the adjoints v_s^* for $s \to O$. It contains terms of positive Grassmann-degree because the pullback metric on $f^* \Omega_Y$ may not coincide with its subspace metric induced from Ω_X. The difference is (in the limit) given by a number-operator $\widetilde{N}$, and $\widetilde{v}^*$ differs from the naive v^* by a factor $\exp(\widetilde{N})$.

We are mostly interested in in the second case of superconnections, but to prove something for it we consider as usual the first and let the metric on Y degenerate. So let $K^\bullet$ denote the mapping cone of $v \colon F \to f_*(E \otimes \wedge^* \Omega_{X/Y}^{0,1})$, and $\widetilde{\nabla}_K = \nabla_F + \nabla_E + v + \widetilde{v}^*$ its superconnection. We are interested in what happens if we replace the metric g_X by $t^{-1} g_X$, scale the metric on $F = K^{-1}$ by $t^{\dim(X)-\dim(Y)-1}$, and let t approach zero or infinity. The power $t^{\dim(X)-\dim(Y)}$ has been put in to account for the scaling of volume-forms. By a perturbation argument one finds that $\mathrm{tr}_s(\exp(-\widetilde{\nabla}_{K,t}^2))$ has an asymptotic expansion as $t \to 0$, and similar for $\mathrm{tr}_s(\widetilde{N} \exp(-\widetilde{\nabla}^2_{K,t}))$, where $\widetilde{N}$ is the number-operator corresponding to a change of g_X. If we replace v by 0 these asymptotic expansions are the differences between those for X and for Y, which had been computed previously. We shall see that the introduction of v does not change them. (Proposition 5.2)

The behavior at $t \to \infty$ is much simpler: As before everything decays exponentially, as $K^\bullet$ is acylic on each fibre. As an application we can introduce a secondary class $\widetilde{\mathrm{ch}}(K^\bullet)$, which is (with the usual scaling by $(2\pi i)^{-p}$ in degree (p,p)) given by

$$\widetilde{\mathrm{ch}}(K^\bullet) = \Big(\frac{1}{2}\int\limits_0^\infty \mathrm{tr}_s\ \big(\widetilde{N}\exp(-\widetilde{\nabla}^2_{K,t})\big)\ \frac{dt}{t}\Big)^{\mathrm{scaled}}$$

Here $\widetilde{N}$ is the number-operator corresponding to the t-scaling. Then

$\lim\limits_{t\to 0} \frac{\partial\bar{\partial}}{\pi i}\ \widetilde{\mathrm{ch}}(K^\bullet) = \int\limits_{X/Y}\mathrm{Td}(T^{1,0}_{X/Y})\cdot\mathrm{ch}(E) - \mathrm{ch}(F)$, and under a change of the

metric g_X $\widetilde{\mathrm{ch}}(K^\bullet)$

changes by $\int\limits_{X/Y}\widetilde{\mathrm{Td}}(T^{1,0}_{X/Y})\cdot\mathrm{ch}(E),$ modulo $\mathrm{Im}(\partial)+\mathrm{Im}(\bar{\partial})$:

This is true for infinitesimal changes by the previous computations, and follows in general by integration.

COMPLEMENTS ON HEAT KERNELS

Consider the previous situation: $f\colon X \to Y$ is a smooth map of compact Kähler-manifolds, E a holomorphic hermitian bundle on X such that $F = f_*E$ is locally free, and all higher direct images vanish. For the metric $g_{s,t} = t^{-1}(g_X + s^{-1}f^*g_Y)$ we consider the Laplacian $\Delta_{s,t} = t \cdot \Delta_{s,1}$ on the mapping cone of $v\colon \Gamma(Y, F \otimes \wedge^*\Omega^{0,1}_Y) \to \Gamma(X, E \otimes \wedge^*\Omega^{0,1}_X)$. Here the metric on $\Gamma(Y, F\otimes\wedge^*\Omega^{0,1}_Y)$ is scaled by an additional factor $t^{\dim(X)-\dim(Y)}$, to account for the volume- forms. This makes sections of F uniformly square-integrable on the fibres $Z = f^{-1}(y)$ of f.

As an illumination we treat first the absolute case, where Y is a point. Then v and all its derivatives are uniformly trace-class, and so is the difference between Δ and the direct sums of the Laplacians on F and $\Gamma(X, E \otimes \wedge^*\Omega^{0,1}_X)$. Thus we may treat this difference as a perturbation: $\exp(-t\Delta)$ is given by a 2×2-matrix whose entries are an endomorphism of F, an integral operator on X, and a C^∞-map $F \to \Gamma(X, E \otimes \wedge^*\Omega^{0,1}_X)$, respectively its adjoint. As $t \to 0$ these have an asymptotic expansion in t, and v influences only terms with a positive t-power. Thus all asymptotics can be computed from the previous formulas.

For $t \to \infty$ everything decays exponentially, as the complex is acyclic and thus all eigenvalues of Δ are positive.

Now we intend to show that this picture also holds in the relative case. This is not quite as trivial because the higher Grassmann-terms in $\widetilde{\nabla}^2, \widetilde{v}$ and $\widetilde{N}$ scale with negative powers of t. We thus have to go back to the basics. As usual the superconnection is treated as the limiting case $s \to 0$ of ordinary Laplacians. In addition we shall so consider the question of what happens if $t \to \infty$, but st remains bounded and small. Here everthing will decay exponentially.

First of all the heat-kernel $\exp(-\Delta_{s,t})$ is again given by a 2×2-matrix of integral operators, with entries in X or Y. After scaling with $t^{\dim(X)} s^{\dim(Y)}$ respectively $(ts)^{\dim(Y)}$ (to account for volume-forms) the restrictions of the diagonal terms to the diagonals of X respectively Y are C^∞ in all arguments, for $s \geq 0, t \geq 0$:

This follows as before using cutoffs in X and Y, and comparing to Taylor-expansions. One might worry that the commutator between v and the cutoff causes trouble, but its norm is $0(\sqrt{t})$ because of the scaling on v.

Next we need to know that the eigenvalues of $\Delta_{s,1}$ are uniformly bounded below by a fixed $\epsilon > 0$, independent of s. This can be done as follows: First fix a $y \epsilon Y$, and let $Z = f^{-1}(y)$. Identify near Z X with $Z \times T_{Y,y}$, scale coordinates on the second factor by $\sqrt{s}$, and for $s = 0$ we obtain the direct sum of the Laplacian of $T_{y,g}$ and the relative Laplacian in the fibre Z. This relative Laplacian has positive eigenvalues, because the complex on Z is acyclic. It follows by perturbation-theory that for each y we can find a neighborhood U_y and an $\epsilon_y > 0$ such that for any C^∞-section with support h in U_y, and small enough s we have

$$< \Delta_{s,t}(h), h > \geq \epsilon_y < h, h > .$$

By compactness we can find finitely many y such that the U_y cover Y, and can choose ϵ independent of y. Now Sobolev-theory works uniformly in s with the metric $g_{s,1}$: Use finitely many charts obtained by identifying a neighborhood of $Z = f^{-1}(y)$ with an open subset of $Z \times T_{Y,y}$, and scale the second coordinate by $\sqrt{s}$. Especially we have a uniform Garding-inequality for $\Delta_{s,1}$. Apply this as follows:

Choose a small open cover U_i of Y such that for h with support in U_i we have

$< \Delta_{s,t} h, h > \geq \epsilon < h, h >$. Choose C^∞-functions φ_i with support in U_i such that $\sum_1 \varphi_i^2 = 1$. Then the Sobolev-norm of $[\Delta_{s,1}, \varphi_i]$ is $O(\sqrt{s})$. For a C^∞-section h on X we now have

$$< \Delta_{s,1}(h), h > = \sum_i < \varphi_i \Delta_{s,1}(h), \varphi_i \cdot f >$$

$$= \sum_i < \Delta_{s,1}(\varphi_i \cdot h), \varphi_i \cdot h > - \sum_i < [\Delta_{s,1}, \varphi_i](h), \varphi_i h >$$

The first term is $\geq \epsilon < h, h >$, while the second one is bounded by a fixed multiple of $\sqrt{s}(< \Delta_{s,1}(h), h > + < h, h >)$. Thus for s small enough we have

$$< \Delta_{s,1}(h), h > \geq \frac{\epsilon}{2} < h, h >,$$

so all eigenvalues of $\Delta_{s,1}$ are $\geq \frac{\epsilon}{2}$.

It now also follows from Sobolev-estimates that as $t \to \infty$, the kernel of $s^{\dim(Y)} \cdot \exp(-t\Delta_{s,1})$ decays exponentially together with all its s-derivatives. In other words the asymptotic s-expansion of $\exp(-t\Delta_{s,1})$ as well as all remainder term decay exponentially at infinity. At last we want to consider the superconnection $\widetilde{\nabla}_t$ as a limit or ordinary connections. We now consider $\exp(-t\Delta_{s,1})$ as an integral operator on $(f_*(E \otimes \wedge^* \Omega_{X/Y}^{0,1})) \oplus F) \otimes \wedge^* \Omega_Y^{0,1}$. As thus it has an asymptotic expansion in s, by the usual scaling and limit arguments. The expansion start with $s^{-\dim(Y)}$, and counting Clifford-degrees in $C(Y)$ we see as before that the $s^{l-\dim(Y)}$-term has filtration-degree $\leq 2l$. Also its leading term is given by the superconnection $\widetilde{\nabla}_t^2$. Thus we can as before use the limits $s \to 0$ to deduce statements about $\exp(-\widetilde{\nabla}_t^2)$. The same applies to an operator $\alpha^{a,s} \cdot \exp(-t\Delta_{s,1})$, where α is a (p,p)-form on X, and $\alpha^{a,s}$ denotes the action on $S(X)$ with respect to the metric $g_{s,1}$. Finally we consider asymptotic expansions as $t \to 0$, for a fixed s. For this we use a cutoff in Y and scale the Y-coordinates by $\sqrt{t}$. Thus $\Delta_{s,t}$ now becomes a perturbation of $t\Delta_Z + \Delta_{T_{Y,y}}$.

Consider the coefficient of $t^{\dim(Y)} \cdot \exp(-t\Delta_{s,1})$ which acts on $F \otimes S(Y)$. It follows from perturbation-theory that it is C^∞ in $\sqrt{t}$. Furthermore we can count Clifford- degrees, with the usual result that terms with a low t-power also have low degree. Finally in the perturbation-expansion the terms involving v have at least a $\sqrt{t}$ in front of them, but Clifford-degree zero. Thus they do not contribute to the leading terms. Integrating against a (p,p)-form α on Y and letting $s \to 0$ we obtain in a now familiar way that the supertrace of the part of the kernel of $\exp(-\widetilde{\nabla}_t^2)$ which acts on F has on the diagonal of $Y \times Y$ no terms with negative t-powers, and that the t^0-term is independent of v. However we can derive that this holds for the whole kernel.

Proposition 5.2.

$\operatorname{tr}_s\left(\exp(-\widetilde{\nabla}_t^2)\right)$ *has an asymptotic expansion in* t *as* $t \to 0$, *starting with a* t^0-*term which is independent of* v. *Also if* $\widetilde{N}$ *denotes the super-number operator associated to a change of the metric* g_X, *then modulo* $\operatorname{Im}(\partial) + \operatorname{Im}(\bar{\partial})$ $\operatorname{tr}_s\left(\widetilde{N}\exp(-\widetilde{\nabla}_t^2)\right)$ *has an asymptotic expansion, starting with* t^{-1}, *and* t^{-1}- *and* t^0-*terms independent of* v.

Proof. It follow from the previous that we have asymptotic expansions. If we replace v by $s \cdot v$ the coefficients are polynomials in s. It thus suffices that for powers t^l with $l \leq 0$ these coefficients are invariant under $s \cdot d/ds$. Now scaling v corresponds to a number-operator N_Y which is 1 on F and vanishes on $f_*\left(E \otimes S(X/Y)\right)$. By the previous theory of N-operators we have to consider t-powers $t^l, l \leq v$, in $\partial\bar{\partial}\operatorname{tr}_x(N_Y \exp(-\widetilde{\nabla}_t^2))$ respectively the ϵ-derivative of $\operatorname{tr}_s(N_Y \cdot \exp(-\widetilde{\nabla}_t^2))$, where ϵ denotes a parameter describing the change of metric g_X which gives rise to $\widetilde{N}$. However we know that up to $O(t)$ $\operatorname{tr}_s(N_Y \cdot \exp(-\widetilde{\nabla}_t^2))$ is independent of v. We thus may assume the $v = 0$. However then everything is invariant under scaling v, thus the derivatives $s\frac{d}{ds}$ vanish on terms involving powers t^l with $l \leq 0$, and we are done.

DIRECT IMAGES IN ARITHMETIC K-THEORY

Suppose X and Y are regular projective schemes over $\mathbb{Z}$, $f : X \to Y$ a flat map which is smooth over $\mathbb{Q}$. We intend to define a direct image $f_* : \hat{K}(X) \to \hat{K}(Y)$, as follows:
Suppose first that E is a vectorbundle on X, with hermitian metric on $X_{\mathbb{C}}$, such that the higher cohomology of E on each fibre of f vanishes. Then $F = f_*(E)$ is a vectorbundle on Y, which we endow with some hermitian metric. Over $\mathbb{C}$ we have associated to the map $v : F \to f_*\left(E \otimes \wedge^* \Omega_{X/Y}^{0,1}\right)$ a secondary class $\widetilde{\operatorname{ch}}(\operatorname{cone}(v))$. We define $f_*[E]$ as the sum of $[F]$ and $\widetilde{\operatorname{ch}}(\operatorname{cone}(v))$. We extend f_* to differential forms by the rule $f_*(\alpha) = \int_{X/Y} \alpha \cdot \operatorname{Td}(T_{X/Y})$. This commutes with base change $Y' \to Y$.

Theorem 5.3.

    a) $f_*[E]$ *does not depend on the metric on* F

    b) f_* *commutes with the forgetful map to* $K(X)$ *and the curvature-homomorphism* h *to* $A'(X)$

    c) *If* $E : 0 \to E_1 \to E_2 \to E_3 \to 0$ *is an exact sequence of acyclic bundles on X, then* $f_*([E_2]) = f_*([E_1]) + f_*([E_3]) + f_*(\widetilde{ch}(E))$.

Proof. b) follows from the curvature computations. For a) the assertion already holds up to a class in the image of $\mathcal{H}(Y)$. Use a metric on the pullback of F to $Y \times \mathbb{P}^1$ which restricts to the two given metrics on $Y \times \{0\}$ respectively $Y \times \{\infty\}$. Then any class in $\mathcal{H}(Y \times \mathbb{P}^1)$ has the same restriction to $\{0\}$ as to $\{\infty\}$, so we are done.

The same argument works for c), deforming E to a split exact sequence. For a general bundle E on X we can find a resolution

$$0 \to E \to E_0 \to E_1 \to \quad \to E_n \to 0,$$

with all E_i acylic. Define $f_*([E])$ as the sum of $\sum(-1)^n f_*([E_n])$ and the direct image of the $\widetilde{ch}$-class of this resolution. This is independent of the choices: Two resolutions map to a common third one, and a map of resolutions can be deformed to a split quasi-isomorphism. Also f_* is independent of g_Y. If we change the metric g_X f_* changes by a $\widetilde{Td}$-class. Next we consider compositions: Suppose we have another map $g\colon Y \to Z$. It is not true that $(g \circ f)_* = g_* \circ f_*$, because on the curvature side we integrate once against $\mathrm{Td}(T_{X/Z})$ and on the other side against $\mathrm{Td}(T_{X/Y}) \cdot \mathrm{Td}(T_{Y/Z})$. This is because in the definition of superconnections we once let the metric on X scale to infinity, which is different from first blowing up the metric on Y and then the metric on X. If we blow up the metric on Y the exact sequence

$$0 \to T_{X/Y} \to T_{X/Z} \to f^* T_{Y/Z} \to 0$$

deformes to the split sequence. However we still can show that the now obvious guess is true. Let $\widetilde{Td}(X, Y, Z)$ denote the secondary class associated to this exact sequence.

Theorem 5.4.

$$(g \circ f)_*([E]) = g_*(f_*([E]) + \epsilon(\int_{X/Y} ch'(E)\widetilde{Td}(X, Y, Z)))$$

Proof. We easily reduce to the case where E is acyclic on the fibres of f and $F = f_*(E)$ acyclic on the fibres of g. Let $G = g_*(F) = (g \circ f)_*(E)$. On $Z_\mathbb{C}$ we have maps

$$G \xrightarrow{v} g_*\left(F \otimes \wedge^* \Omega^{0,1}_{Y/Z}\right) \xrightarrow{u} (g \circ f)_*\left(E \otimes \wedge^* \Omega^{0,1}_{X/Z}\right)$$

By definition $(g \circ f)_*([E]) = [G] + \widetilde{ch}(\text{cone}(u \circ v))$. We remark that $\text{cone}(v), \text{cone}(u \circ v)$ and $\text{cone}(u)$ have superconnections, and their associated heat-kernels $\exp(-\widetilde{\nabla}_t^2)$ have asymptotic expansions as $t \to 0$: This follows by the usual cutoff-procedure. We can thus define their $\widetilde{ch}$-classes. We need

Lemma 5.5.

$$\widetilde{ch}(\text{cone}(u \circ v)) = \widetilde{ch}(\text{cone}(u)) + \widetilde{ch}(\text{cone}(v))$$

Proof. $(1, u)$ induces a map $w: \text{cone}(v) \to \text{cone}(u \circ v)$. As before $\text{cone}(w)$ has a superconnection and an asymptotic expansion for $\exp(-\widetilde{\nabla}_t^2)$ as $t \to 0$. We claim that up to positive t-powers $\text{tr}_s\left(\exp(-\widetilde{\nabla}_t^2)\right)$ is independent of w and v, i.e. it is the same as for the cone of the zero-map $\text{cone}(0) \to \text{cone}(0)$: First the previous arguments give that it is independent of scaling in G, by which we can make v equal to zero. Then w becomes the direct sum of the identity map on G and the map u. For both of them independence of scaling follows from the usual count of Clifford-degrees. Now over $Z \times \mathbb{P}^1$ we can deform v to zero (considering $z \cdot v: G \to g_*(F \otimes \wedge^*\Omega_{Y/Z}^{0,1})(1)$). The curvature of the class $\widetilde{ch}(\text{cone}(w))$ is independent of z, thus the restriction of the $\widetilde{ch}$-class to $Z \times \{0\}$ and $Z \times \{\infty\}$ coincide. By the same argument we also can deform w to zero. But if $v = 0$ the $\widetilde{ch}$-class is equal to $\widetilde{ch}(\text{cone}(u)) - \widetilde{ch}(\text{cone}(1_G)) = \widetilde{ch}(\text{cone}(u))$, where for $w = 0$ we obtain $\widetilde{ch}(\text{cone}(u \circ v)) - \widetilde{ch}(\text{cone}(v))$. Thus the assertion follows. (Note that all the complexes involved are acylic, so the $\widetilde{ch}$-classes exist).

Now using the lemma, changing the metric g_X to $g_X + s^{-1}f^*g_Y$ and letting $s \to 0$ (which kills the $\widetilde{Td}$-class) the asertion of the theorem amounts to checking that the limit of $\widetilde{ch}(\text{cone}(v))$ is equal to $\int_{X/Z} \widetilde{ch}(\widetilde{v})\,\text{Td}(T_{Y/Z})$, where $\widetilde{v}: F \to f_*(E \otimes \wedge^*\Omega_{X/Y}^{0,1}))$ denotes the augmentation on Y. Both sides of the hoped for equality are given by regularized integrals

$$\int_0^\infty \text{tr}_s\left(\widetilde{N} \cdot \exp(-\widetilde{\nabla}_t^2)\right)\frac{dt}{t},$$

where t corresponds to scaling g_X to $t^{-1} \cdot g_X$ respectively $g_X + s^{-1}f^*(g_Y)$ to $t^{-1}g_X + (st)^{-1}f^*(g_Y))$, and $\widetilde{N}$ is the corresponding number-operator. If in the plane we parametrise the x-axis by t and the y-axis by (st), the integral is over the straight line $y = sx, 0 < x < \infty$. By Theorem 5.1. we can deform the path of integration as follows:

First go from $(0,0)$ to $(\epsilon, s\epsilon)$ to ϵ small. Then from $(\epsilon, s\epsilon)$ to $(\epsilon, t_0\epsilon), t_0$ big. Finally vertically from $(\epsilon, t_0\epsilon)$ to $(st_0 \epsilon, t_0\epsilon)$, and from there to infinity as before. If we first let $t_0 \to \infty$ the last integrals decay exponentially. So we can also first move from $(0,0)$ to $(\epsilon, s\epsilon)$, then parallel to the x-axis to $(\infty, s\epsilon)$. For the latter integral the number-operator corresponds to scaling gx. It thus converges for $\epsilon \to 0$ to the super-number-operator used in the definition of $\tilde{v}$. So finally we deform to the following path:

0_ϵ: From $(0,0)$ in a straight line to $(\epsilon, s\epsilon)$.

I_ϵ: From $(\epsilon, s\epsilon)$ vertically down to $(\epsilon, 0)$

II_ϵ: Along the x-axis from $(\epsilon, 0)$ to $(\infty, 0)$.

Now at the origin all integrands have asymptotic expansions in s and t and thus also in x and y. Furthermore as the integral over I_ϵ exists (the others do) its integrand has an asymptotic expansion with no negative y-powers. It follows that the integral over I_ϵ converges if $s \to 0$, for ϵ fixed. It follows that for any $\epsilon > 0$ the difference between the two $\tilde{\text{ch}}$-classes is given by the limit (as $s \to 0$) of the difference of regularized integrals over 0_ϵ and $\tilde{0}_\epsilon$, where $\tilde{0}_\epsilon$ denotes the path from $(0,0)$ to $(\epsilon, 0)$ with the integrand given by the trace of the number-operator for $\tilde{v}$. Expecially the limit exists. We thus have the following situation:

The difference of integrands is a function $f(s,t)$ which as $t \to 0$ has an asymptotic expansion

$$f(s,t) \sim \sum_i a_i(s) t^i.$$

We know that the limit $\lim_{s \to 0} \int_0^\epsilon f(s,t) \, dt$ exists, and is independent of ϵ.

Differentiating with respect to ϵ we derive that the $a_i(s)$ are regular as $s \to 0$, and $a_i(0) = 0$. Thus $\lim_{s \to 0} \int_0^\epsilon f(s,t) \, dt = 0$, and we are done.

LECTURE 6. ARITHMETIC RIEMANN–ROCH THEOREM

In this lecture we intend to prove a Riemann-Roch theorem for arithmetic varieties. After some general discussions we prove the Riemann-Roch theorem for P^1 bundles with a modified Todd class. Then we try to reduce the general Riemann-Roch to some deformation argument to the normal cone, for closed embeddings. Finally we prove the deformation theorem by some detailed estimates on heat kernels. References [BL], [F], [GS3], [GS4]

GENERALITIES

We want to prove a Riemann-Roch theorem for a smooth and proper $f: X \to Y$, where X and Y are regular projective schemes. It has been shown by Gillet-Soulé ([GS3]) that for this we have to add a secondary class to the Todd-genus. For any power series $R \epsilon \mathbb{R}[[x]]$ we have an associated additive characteristic class $R(E)$. Define

$$\widehat{\mathrm{Td}}^R(E) = \widehat{\mathrm{Td}}(E)\left(1 - \epsilon(R(E))\right)$$

We then shall show:

Theorem 6.1.

There exists a unique power-series $R(X)$ such that

$$\widehat{\mathrm{ch}}(F_*([E])) = f_*\left(\widehat{\mathrm{ch}}(E) \cdot \widehat{\mathrm{Td}}^R(T_{X/Y})\right)$$

Remark. We shall not try to compute R. This has been done by Gillet-Soulé, by evaluating for $X = \mathbb{P}^n, Y = \text{point}, E = O_X$. They get

$$R(X) = \sum_{m \text{ odd}, \geq 1} [2\zeta'(-m) + \zeta(-m)(1 + \frac{1}{2} + \frac{1}{3} + \cdots + \frac{1}{m})]X^m/m!$$

Here ζ denotes the Riemann-zetafunction. The even terms can be determined by considering the effect of Serre-duality for sheaves on $\mathbb{P}^n_{\mathbb{Z}}$.

Before we proceed some more general remarks:

    i) The assertion holds after composing with the forgetful-map to $A(X)$ or the curvature-homomorphism h. That is the difference of the two sides lies in $\epsilon\big(\mathcal{H}(X)\big)$.

ii) The assertion is independent of the Kähler-metrics on X and Y, or the hermitian metrics on E. Thus the error is given by a map

$$\mathrm{Err}_f\colon K(X) \longrightarrow \epsilon\big(\mathcal{H}(Y)\big).$$

iii) The assertion is compatible with compositions:

$$\mathrm{Err}_{g\circ f}(E) = g_*\big(\mathrm{Err}_f(E)\big) + \mathrm{Err}_g\big(f_*(E)\big)$$

iv) Everything is compatible with base-change.

v) Err is $K(Y)$-linear.

Riemann-Roch theorem for $\mathbb{P}^1$-bundles.

Theorem 6.2. *There is an unique power series R such that the Riemann-Roch formula holds for $\mathbb{P}^1$-bundles: $f : X = \mathbb{P}_Y(E) \to Y$.*

Proof. By the remarks above it suffices to prove that there is an unique R such that Riemann-Roch formula holds for O_X and $O_X(-1)$, and some metric on T_X.

We take a metric on E and put the quotient metric on $O(1)$. From the exact sequence

$$0 \to O \to E^\vee \otimes O(1) \to T_{X/Y} \to 0$$

we have that $T_{X/Y} = (\det E)^\vee \otimes O(2)$. We put a metric on $T_{Y/X}$ such that is an isometric map.

Let $h' = \frac{1}{2}c_1'(T_{X/Y})$ then $h'^2 = -c_2'(E) + \frac{1}{4}c_1'(E)^2$. This implies that $f_* h'^{2n} = 0$ and

$$f_* h'^{2n+1} = (h'^2)^n f_* h' = (h'^2)^n.$$

Now let $Y_n = \mathrm{Grass}(2,n)$ denote the Grassmannian of rank-2-quotients of $\mathbb{Z}^{n+2}$, and $X_n = \mathbb{P}(E_n) \to Y_n$ the projective bundle associated to the universal E_n. For $E = O_X$ or $O_X(-1)$ the Riemann-Roch holds up to an error in $\mathcal{H}^*(Y_n)$ (there ϵ is an injection, as this holds for Spec $(\mathbb{Z})$). These errors correspond under the injections $Y_n \hookrightarrow Y_{n+1}$, so we obtain universal classes in $\varprojlim \mathcal{H}(Y_n) = \mathbb{R}[[c_1'(E), c_2'(E)]]$. They have the property that for any rank-2-bundle E which is generated by its global section, these classes define the error in Riemann-Roch for $\mathbb{P}_Y(E) \to Y$. Expecially as this error remains unchanged if we tensor E_n with a line-bundle, we see that the error for O_X must be given by a power-series in $c_1'^2(E) - 4\,c_2'(E)$ (tensor the E_n with powers of an ample bundle on Y_n). The same is true for $O(-1)$ if we multiply the error by $\exp\left(\frac{1}{2}c'_1(E)\right)$. Now we have to check that we can adjust R uniquely such that both errors become zero.

If we change R by δR than the Riemann-Roch expression for O_X changes by

$$f_*\left(\frac{2h'}{1 - e^{-2h'}} \cdot \delta R(2h')\right)$$

and for $O(-1)$ $\left(\text{up to } \exp\left(\frac{1}{2}c'_1(E)\right)\right)$ by

$$f_*\left(\frac{2h' \cdot e^{-h'}}{1 - e^{-2h'}} \delta R(2h')\right)$$

In the second expression the series before δR is even. We thus can choose the odd part δR^{odd} uniquely such that the error vanishes for $O(-1)$. So now assume that δR is even, and try to cancel the error for O_X. Here only the odd-part in the power-series matters, that is we have to consider

$$f_*(h' \cdot \delta R(2h')).$$

Again there is a unique δR.

So we have shown Riemann-Roch for $\mathbb{P}^1$-bundles $\mathbb{P}(E)$, if E is globally generated. But for a projective Y we can always achieve this by tensoring with an ample line-bundle.

DEFORMATION TO THE NORMAL CONE

Deformation theorem.

We fix an arithmetic variety B and consider only arithmetic varieties which are smooth over B. As usual we define for such an arithmetic variety and a hermitian vector bundle E

$$\mathrm{Err}_X(E) = \widehat{\mathrm{ch}}\, f_*[E] - \int_{X/B} \widehat{\mathrm{ch}}(E) \cdot \widehat{\mathrm{Td}}^R(E)$$

where R is the unique power series defined in the last section. Let $i : Y \to X$ be an immersion. Let $\mathrm{Err}_i(E)$ denote $\mathrm{Err}_X(i_*E) - \mathrm{Err}_Y(E)$ for any vector bundle E on Y.

Let $\widetilde{X}$ denote the blow-up of $X \times \mathbb{P}^1$ in $Y \times \{\infty\}$. So $\widetilde{X} \to X \times \mathbb{P}^1$ is a birational map such that over $\mathbb{P}^1 - \{\infty\} = \mathbb{A}^1$ it is an isomorphism and X_∞ is an union of two smooth irreducible components X'_∞ and X''_∞ which meet transversally. X''_∞ is isomorphic to the blow-up of Y in X, while X'_∞

is a projective bundle over Y, as is $\partial X_\infty = X'_\infty \cap X''_\infty$, which is also the exceptional divisor in X''_∞. We have a map $\tilde{Y} = Y \times \mathbb{P}^1 \to \tilde{X}$ which lifts the canonical embedding $Y \times \mathbb{P}^1 \to X \times \mathbb{P}^1$ such that $\tilde{Y}$ is disjoint from X''_∞.

Assume now that Y has codimension one in X. Let $\tilde{I}_Y$ denote the ideal of $\tilde{Y} \subset \tilde{X}$. We put on $\tilde{I}_Y$ a metric that $\tilde{I}_Y = O_{\tilde{X}}$ (isometric) in a neighborhood U of X''_∞ which does not meet $\tilde{Y}$. Then we only consider the formal difference $[\tilde{E}] - [\tilde{I}_Y \cdot \tilde{E}]$, $\tilde{E}$ a vector bundle on $\tilde{X}$.

We shall choose $\tilde{E} = E(X''_\infty)$, where E is (the pullback to $\tilde{X}$ from) a hermitian bundle on X such that E and $I_Y \cdot E$ are acylic relative B. We denote by F^{-1} (respectively F^0) the direct image of $I_Y E$ (respectively E) on B. Those are bundles which we endow with some hermitian metric. Finally $F^\bullet$ denotes the comples $F^{-1} \hookrightarrow F^0$ on B, and $E^\bullet = [E^{-1} \to E^0]$ the corresponding complex $I_{\tilde{Y}} \cdot E(X''_\infty) \hookrightarrow E(X''_\infty)$ on $\tilde{X}$. On $B \times \mathbb{P}^1$ we have a quasiisomorphism $v \colon F^\bullet \to f_*(E^\bullet)$, for each t a $v_t = F^\bullet \to f_*(E^\bullet|X_t)$, as well as $v_\infty \colon F^\bullet \to f_*(E^\bullet|X'_\infty)$.

Theorem 6.3. *Suppose $Y \subset X$ is a divisor. With any Kähler-metric on $\tilde{X}$*

$$\lim_{t \to \infty} \tilde{\mathrm{ch}}_t = \tilde{\mathrm{ch}}_\infty,$$

where $\tilde{\mathrm{ch}}_t$ denotes the ch -class for $\mathrm{cone}(v_t)$ that is the regularized integral $\int_0^\infty \mathrm{Tr}_s(\tilde{N} e^{\tilde{\nabla}_u^2}) \frac{du}{u}$ as usual.

We claim that this implies the following approximation to Riemann-Roch:

Theorem 6.4. *The error in Riemann-Roch for i_0 and $i^*(E)$ (that is the difference of errors for $(Y, E|_Y)$ and $(X, i_* i^* E)$) is the same as that for i_∞ and $i_\infty^* E$.*

Proof. On the RHS of Riemann-Roch we have classes $\int \hat{\mathrm{ch}} \cdot \mathrm{Td}^R$. These concide on Y. On X we first replace the tangent bundle T_{X_0} resp $T_{X'_\infty}$ by the restriction of the logarithmic tangent bundle $T = T_{\tilde{X}/\mathbb{P}^1}$. We put on T a metric such that outside U the isomorphism $T_t = T_{X_t}$ is an isometry. Then one shows that in the fiber of 0 or ∞ we have that

$$\hat{\mathrm{ch}}(\tilde{E}^\bullet) \cdot \widehat{\mathrm{Td}}^R(T_{X_t}) = \hat{\mathrm{ch}}(\tilde{E}^\bullet) \cdot \widehat{\mathrm{Td}}^R(T)$$

Use that $\hat{\mathrm{ch}}(\tilde{E}^\bullet)$ is of the form (Z, g_Z), where Z is a cycle supported in $\tilde{Y}$ and g_Z a Green's function supported in $W = \tilde{X} - \overline{U}$.

In fact define a group $\hat{A}_{Y,W}(X)$ as the quotient of cycles (Z, g_Z) with $Z \subseteq Y$, $\text{supp}(g_Z) \subseteq W$, modulo divisors of rational functions on cycles in Y, and forms $\partial\alpha + \bar{\partial}\beta$ where α and β are currents with support in Y. One checks that for any hermitian line-bundle L on X $\hat{c}_1(L)$ still acts on $\hat{A}^*_{Y,W}(X)$, and two $\hat{c}_1(L)$'s commute. Furthermore if we have two embeddings $Y \hookrightarrow X_1$ and $Y \hookrightarrow X_2$ such that there are Zariski-open neighborhoods V_1 respectively V_2 of Y and an isomorphism $V_1 \cong V_2$ respecting Y, and if $W_1 \subseteq V_{1,\mathbb{C}}$ and $W_2 \subseteq V_{2,\mathbb{C}}$ are corresponding open neighborhoods, then naturally $\hat{A}^*_{Y,W_1}(X_1) \cong \hat{A}^*_{Y,W_2}(X_2)$. Furthermore this isomorphism respects the operation of $\hat{c}_1(L)$'s, if there is an isomorphisms $L_1|V_1 \cong L_2|V_2$ which is an isometry over $W_1 \cong W_2$.

Now suppose that we have two hermitian vectorbundles E_1 and E_2 which are isomorphic over a Zariski-open $V \subseteq Y$, isometric over $W \subseteq V_\mathbb{C}$. Let (Z, g_Z) represent a cycle in $A^*_{(Y,W)}(X)$. If X_1 (respectively X_2) denotes the complete flag-varietites of E_1 (respectively E_2), $\pi_{1/2}: X_{1/2} \to X$ the projections, $V_{1/2} = \pi_{1/2}^{-1}(V)$ etc., then we have corresponding cycles $\pi^*_{1/2}(Z, g_Z)$ in $\hat{A}^*_{(Y_{1/2}, W_{1/2})}(X_{1/2})$.

Now on $X_{1/2}$ $\pi^*_{1/2}(E_{1/2})$ has a complete filtration by line-bundles L_j. Thus we can let any polynomial in the $\hat{c}_1(L_j)$ operator on our cycle, and the two results correspond. The same holds for multiplication by $\tilde{\text{ch}}$-classes. Now suppose $P(\hat{c}_i(E))$ is a polynomial in the chern-classes of E. Then we find an operator $Q((\hat{c}_1(L_j)) + R$ on $\hat{A}^*_{1/2}(X_{1/2})$, where Q is a polynomial in Chern-classes and R multiplication by a secondary class, such that in $\hat{A}^*(X)$.

$$P(\hat{c}_i(E_{1/2}))(Z_1, g_Z) = \pi_{1/2*}((Q(\hat{c}_1(L_j))) + R) \cdot \pi^*_{1/2}(Z, g_Z))$$

It follows that both for E_1 and E_2 the left-hand sides can be obtained from the same class in $\hat{A}^*_{(Y_{1/2}, W_{1/2})}(X_{1/2})$, by first projecting to $\hat{A}^*_{(Y,W)}(X)$ and then mapping to $\hat{A}^*(X)$. Thus the operations of $P(\hat{c}_i(E_{1/2}))$ on cycles with support in (Y, W) coincide.

In the Riemann-Roch for $E^\bullet$ the difference in RHS's is

$$\widehat{\text{ch}}(E^\bullet) \widehat{\text{Td}}^R(T) \cdot (0 - \infty, 0) = (0, \int_{\mathbb{P}^1} \text{ch}'(E^\bullet) \cdot \text{Td}'(T) \cdot \log|z|),$$

using that

$$\text{div}(z) = (0 - \infty, -\log|z|) = 0.$$

This is the limit:

$$\lim_{t \to \infty} (0, \int_{\mathbb{P}^1} \text{ch}'(E^\bullet) \cdot \text{Td}'(T) \cdot \log|\frac{zt}{z-t}|),$$

which is the limit of secondary classes describing the change of metrics $(0 \to t)$ on $E^{\bullet}|_{X_t}$ and $T|_{X_t}$. On the LHS we have $\widehat{\mathrm{ch}}(F^{\bullet}) + (0, \widetilde{\mathrm{ch}}(\text{class}))$. The difference between 0 and ∞ is

$$\left(0, \int_{\mathbb{P}^1} \mathrm{ch}'(F^{\bullet}) \cdot \log|z|\right) + (0, \widetilde{\mathrm{ch}}_{\infty} - \widetilde{\mathrm{ch}}_0).$$

This is the limit of secondary classe where we integrate $\log\left|\frac{zt}{z-t}\right|$, which gives the secondary class for base change $0 \to t$ as before. As Riemann-Roch is compatible with these secondary classes, we derive the assertion. We also have to use some standard arguments relating the $\widetilde{\mathrm{ch}}(v_t)$ to the previous $\widetilde{\mathrm{ch}}$-classes. This is done by $\mathbb{P}^1$-deformations.

We thus can replace i_0 by i_∞ in the discussion of Riemann-Roch.

Situations.

Let us first deform any $Y \to X$ to its normal cone $\mathbb{P}_Y(O_Y \oplus I_Y/I_Y^2)$. Let $\widetilde{X} =$ blow-up of $X \times \mathbb{P}^1$ in $Y \times \infty$, $X_0 =$ fiber of $\widetilde{X}$ at $0 = X \times 0$, $X_\infty =$ fiber of $\widetilde{X}$ at $\infty = X_\infty' \cup X_\infty''$, $X_\infty'' =$ blow-up of X in Y, $X_\infty' = \mathbb{P}_Y(O_Y \oplus I_Y/I_Y^2)$, $\widetilde{Y} = Y \times \mathbb{P}^1 \subset \widetilde{X}$, $Y_\infty = Y \subset X_\infty'$.

If F is any sheaf on Y, $\widetilde{F} = pr_1^*(F)$ on $\widetilde{Y}$, then $\widetilde{F}/X_\infty$ and $\widetilde{F}/X_0$ respresent the same element in K-theory, so the compositions

$$Y \overset{i_\infty}{\hookrightarrow} X_\infty' \overset{j_\infty}{\hookrightarrow} \widetilde{X}$$

$$Y \overset{i_0}{\hookrightarrow} X_0 \overset{j_0}{\hookrightarrow} \widetilde{X}$$

induce the same map in K theory , or on cycles.

Thus the Riemann-Roch holds for $(F$ and $i_0)$ if it holds for $(F$ and $i_\infty)$, $(i_{\infty,*}(F)$ and $j_\infty)$ and $(i_{0,*}(F)$ and $j_0)$. Note that for the second and third immersion the bundle is induced from $\widetilde{F}$, so from the big scheme. The same is true in the first case, as i_∞ is the zero-section in a projective bundle. To treat general immersions it thus suffices to show that Riemann- Roch holds for immersions $i\colon Y \hookrightarrow X$ which are either of codimersion one, or such that $X = \mathbb{P}_Y(O_Y \oplus N)$ is a projectivised affine bundle, and i the zero-section. Also it suffices to treat bundles $F = i^*(E)$ which are pullbacks from X. As usual we may assume that E is acyclic relative B. Finally for the immersion $i\colon Y \hookrightarrow X = \mathbb{P}_Y(O_Y \oplus N)$ the Riemann-Roch holds for i if it holds for the projection $\pi\colon X \to Y$, as $\pi o i = id$. It now follows from Theorem 6.4 that the Riemann-Roch holds for immersions of codimension one, as it holds for $\mathbb{P}^1$-bundles by Theorem 6.2.

Next we treat the case of a $\mathbb{P}^n$-bundle $X = \mathbb{P}_B(E)$. We use induction over the rank of E, starting with rank 2. The map $\mathrm{Flag}(E) \to X$ thus already satisfies Riemann-Roch, and by the Bott trick Riemann-Roch holds for $O(-1)$ on $\text{Flag}(E)$ and thus also on X: Factor $\text{Flag}(E) \to B$ through a $\mathbb{P}^1$-Bundle $\text{Flag}(E) \to X'$ such that $O(-1)$ has trivial direct image.

Finally by the splitting principle we may assume that E contains a line-bundle L so that is: $Y = \mathbb{P}(E/L) \hookrightarrow X$ is a closed immersion of codimersion one. By induction the Riemann-Roch holds for $i_* K(Y)$, which together with $O(-1)$ generates $K(X)$ over $K(B)$.

This settles the case for projective bundles and thus for arbitrary closed immersions. Finally any projective morphism can be factored into such, so we are done and Theorem 6.1 has been shown.

GENERALIZATIONS AND COMMENTS

First we remark that we can generalize the Riemann-Roch slightly: Suppose $f \colon X \to Y$ is a flat morphism of regular projective schemes (over $\mathbb{Z}$) such that $f_{\mathbb{Q}}$ is smooth. As before we can define a direct image $f_* \colon \hat{K}(X) \to \hat{K}(Y)$ such that for an E on X which is acylic on the fibres, $f_*([E])$ is given by $[f_*(E)] + \tilde{\text{ch}}$-class of $\left(\text{cone } f_*(E) \to f_*(E \otimes \wedge^* \Omega_{X/Y}^{0,1}) \right)$. Furthermore we have a relative Todd-class $\widehat{\text{Td}}_{X/Y}$, defined as follows: Factor f as $f = g \circ i$, where $i \colon X \hookrightarrow P = \mathbb{P}^n \times Y$ is a closed immersion, and g is the projection from P to Y. If $N = (I_X/I_X^2)^*$ denotes the normal bundle to i, we have a map $i^* T_{P/Y} \to N$ which over $\mathbb{Q}$ is surjective, with kernel $T_{X/Y}$. Now endow $T_{P/Y}$ and N with hermitian metrics, and define $\text{Td}_{X/Y}$ is the quotient $\widehat{\text{Td}}(i^* T_{P/Y})/\widehat{\text{Td}}(N)$, modified by the secondary Todd-class associated to $0 \to T_{X/Y} \to i^* T_{P/Y} \to N \to O$ (over $X_{\mathbb{C}}$). One checks that this is independent of all choices.

Theorem 6.5. *Let $x \in \hat{K}(X)$. Then in $\hat{A}(Y)$ we have*

$$\widehat{\text{ch}}\left(f_*(x)\right) = f_*\left(\widehat{\text{ch}}(x) \cdot \widehat{\text{Td}}_{X/Y}^R\right)$$

Proof. As before the assertion holds up to a class in $\mathcal{H}(Y)$. Furthermore we may use the same deformation arguments as before to reduce to the case of smooth projective bundles, for which the result holds already.

We also want to comment on the case where $Y = \text{Spec}(O_K)$ is the ring of integers of a numberfield K. Then $A^1(Y) = Pic(O_K) = $ divisor-class group of K, and the kernel of $\hat{A}_1(Y) \to A^1(Y)$ is the quotient of $\mathbb{R}^{s+t}$ ($s + t = $ number of infinite valuations of K) by the logarithms of units of O_K^*. Furthermore $K(Y)$ is generated by projective O_K-modules P which have hermitian metrics at the infinite places. These define volume forms on $\lambda(P) = \det(P)$. If P is an acylic complex of such modules its determinant $\lambda(P) = \otimes\lambda(P^n)^{\pm 1}$ is canonically isomorphic to O_K. This isomorphism has norm $\exp(\lambda_v)$ at each infinite place v of K, and the secondary class $\widehat{ch}(E^*)$ is given by the λ_v. The normation is that if $E^*\colon E^0 \xrightarrow{\alpha} E^1$ is an isomorphism such that $\lambda(\alpha)$ has norms $\exp(\lambda_v)$, then $\widehat{ch}(E^0) - \widehat{ch}(E^1)$ is represented by the λ_v.

Now suppose we have a bundle E on X such that E is acyclic on the fibres of f, and let $F = f_*E$ by represented by $P = \Gamma(X, E)$. Endow P with the metric given by L^2-integration. Then the direct image $f_*([E])$ is represented by the sum of $[P]$ and the $\widetilde{ch}$-class of the cone of the inclusion $\alpha\colon P_{\mathbb{C}} \to \Gamma\big(X_{\mathbb{C}}, E \otimes S(X)\big)$, which in turn is half the regularized integral

$$-\tfrac{1}{2} \int_0^\infty \mathrm{tr}_s(N\, e^{-t\Delta})\tfrac{dt}{t}.$$

As complex of Hilbert-spaces $\mathrm{cone}(\alpha)$ splits into the orthogonal direct sum of the isomorphism $P_{\mathbb{C}} \xrightarrow{\sim}$ harmonic 0-forms, and the isomorphism give by $\bar\partial \ker(\Delta)^\perp \xrightarrow{\sim} \overline{\mathrm{Im}(\bar\partial)}$, where the left is a subspace of $\Gamma(X, E \otimes \Omega_X^{0,q-1})$ and the right of $\Gamma(X, E \otimes \Omega_X^{0,q})$, for $1 \le q \le \dim(X_{\mathbb{C}})$. Under these isomorphisms the Δ-operators correspond. If $\{\lambda_{q,n} | n = 1, 2, \dots\}$ denotes the eigenvalues of Δ on these spaces $(q = 1, 2, \dots)$, then the zeta-function

$$\zeta(s) = \frac{1}{\Gamma(s)} \cdot \frac{1}{2} \int_0^\infty \mathrm{tr}_s(N\, e^{-t\Delta})\frac{dt}{t}$$

is equal to $\tfrac{1}{2}\sum\limits_{q,n}(-1)^q \cdot \lambda_q^{-s}$, so formally its derivative at 0 is $\zeta'(0) = \log\big(\prod\limits_{q,n} \lambda_{q,n}^{\frac{1}{2}\cdot(-1)^{q-1}}\big)$.

As $\lambda_{q,n}^{\frac{1}{2}}$ are the norms of $\bar\partial$ on the various eigenspaces, the right-hand side is also called the logarithm of the determinant of the $\bar\partial$-operator.

A similar statement even holds if E is not acyclic:
Over the complex numbers the cohomology $H^*(X, E) \otimes \mathbb{C}$ is isomorphic to the space of Δ-harmonic forms $H^* \subseteq \Gamma(X, E \otimes \wedge^*\Omega_X^{0,1})$. If we represent $H^*(X, E)$ by a finite complex $P^\bullet$ of projective O_K-modules, and endow each P^n with a hermitian metric, then $\lambda(P_{\mathbb{C}}) \cong \lambda(H^*)$. The norms at the infinite places of this canonical isomorphism give a secondary class. Another one is obtained as $\widetilde{ch}$ of the cone of $H^* \to \Gamma(X, E \otimes \wedge^*\Omega_X^{0,1})$, and is equal to the determinant of $\bar\partial$ on $H^{*,\perp}$. We claim that $f_*([E])$ is represented by the alternating sum $\sum(-1)^n[P^n]$, plus the sum of these two secondary classes:
By definition $f_*([E])$ is computed by resolving E by acylic bundles, i.e. we

have a quasi-isomorphism $\alpha\colon E \to E^{\bullet}$ with E^n acylic. Let $F^{\bullet}$ denote the complex $\Gamma(X, E^{\bullet})$ with the L^2-norm, and consider the sequence of maps

$$H^* \xrightarrow{\ i\ } \Gamma\left(X, E \otimes \wedge^* \Omega_X^{0,1}\right) \xrightarrow{\ \alpha^*\ } \Gamma\left(X, E^{\bullet} \otimes \wedge^* \Omega_X^{0,1}\right) \xleftarrow{\ j\ } F^*$$

These exists a map of complexes $H^* \xrightarrow{\ \beta\ } F^*$ such that $j \circ \beta$ is homotopic to $\alpha \circ i$. Also one checks that $\widetilde{\mathrm{ch}}\left(\mathrm{cone}(\alpha^*)\right)$ is equal to $\int_{X_{\mathbb{C}}} \widetilde{\mathrm{ch}}\left(\mathrm{cone}(\alpha)\right) \mathrm{Td}_X$ (deform α to a split sequence) and thus $f_*([E])$ is represented by $[F^*] +$ $\widetilde{\mathrm{ch}}\left(\mathrm{cone}(\alpha^*)\right)$. Also as before one checks that $\widetilde{\mathrm{ch}}\left(\mathrm{cone}(\alpha^*)\right) = \widetilde{\mathrm{ch}}\left(\mathrm{cone}(\alpha^* \circ i)\right) - \widetilde{\mathrm{ch}}\left(\mathrm{cone}(i)\right)$, so this can be rewritten as

$$[F^{\bullet}] + \widetilde{\mathrm{ch}}\left(\mathrm{cone}(j)\right) + \widetilde{\mathrm{ch}}\left(\mathrm{cone}(i)\right) - \widetilde{\mathrm{ch}}\left(\mathrm{cone}(\alpha^* \circ i)\right)$$

Also we shall check below that $\widetilde{\mathrm{ch}}$ is invariant under homotopies. So we may replace $\alpha^* \circ i$ by $j \circ \beta$, and obtain with the same argument that

$$f_*([E]) = [F^{\bullet}] + \widetilde{\mathrm{ch}}\left(\mathrm{cone}(i)\right) - \widetilde{\mathrm{ch}}\left(\mathrm{cone}(\beta)\right).$$

That is we represent $H^*(X, E)$ by the complex $F^{\bullet}$. Thus over $\mathbb{C}$ the cohomology $H^*(F_{\mathbb{C}})$ is isomorphic to H^*, via β. Combine this with the inclusion $H^* \hookrightarrow \Gamma(X_{\mathbb{C}}, S(X))$. Then $f_*([E])$ is represented by $[F^{\bullet}]$ and the $\widetilde{\mathrm{ch}}$-class of $H^*(F_{\mathbb{C}}^{\bullet}) \xrightarrow{\ \sim\ } H^* \hookrightarrow \Gamma(X_{\mathbb{C}}, S(X))$.

We still have to check the invariance of $\widetilde{\mathrm{ch}}$ under homotopy. This holds for finite dimensional complexes, as $\widetilde{\mathrm{ch}}$ is given by the norm of the induced map on cohomology. If we have two homotopic maps $\alpha, \beta\colon K \to L^{\bullet}$ with $L^{\bullet}$ of infinite dimension, but $K^{\bullet}$ finite dimensional, we construct a complex $M^{\bullet}$ with $M^n = K^n \oplus K^n \oplus K^{n-1}$, two homotopic maps $\alpha', \beta'\colon K^{\bullet} \longrightarrow M^{\bullet}$, and $\vartheta\colon M^{\bullet} \longrightarrow L^{\bullet}$ with $\alpha = \vartheta \circ \alpha', \beta = \vartheta \circ \beta'$. By using the additivity of $\widetilde{\mathrm{ch}}$ we may replace (α, β) by (α', β'), and thus we are done.

Finally suppose that $X \longrightarrow Y = \mathrm{Spec}(O_K)$ is a semistable curve of genus $g \geq 1$. Then X has a canonical hermitian metric, the Arakelov-metric. Furthermore for hermitian bundles L on X, whose curvature satisfies a certain condition, we have previously defined a volume-form on $\lambda\left(H^*(X_{\mathbb{C}}, L)\right)$, unique up to a common constant factor independent of L. (see [F])

If L has degree $g - 1$ the curvature of the metric on $\lambda(L)$ (which is a line-bundle on the Jacobian $J(X)$) has been computed. It coincides with the curvature of the metric defined via the determinant of $\bar{\partial}$. (By our methods this computation is easy. One obtains the direct image (under $X \times J \to J$) of $\mathrm{ch}'(L) \cdot \mathrm{Td}_X'$). Thus these two metrics coincide up to a factor. For line-bundles of arbitrary degree one can use that both volumes satisfy a

Riemann-Roch formula. From this and some approximation argument one derives that the volumes coincide in general.

Estimates: Proof of theorem 6.3

We now supply the necessary analytic details for the proof of Theorem 6.3. Let us recall the situation:

We have a smooth projective $\mathbb{C}$-scheme B as base, a smooth projective B-scheme X and a smooth relative divisor $Y \subseteq X$. In general f always denotes the maps into B (so $f\colon X \to B$ and $f\colon Y \to B$). Furthermore on X we have a hermitian vectorbundle E such that E and $I_Y \cdot E$ are acylic relative B. $\tilde{X}$ denotes the blow-up of $X \times \mathbb{P}^1$ in $Y \times \{\infty\}$. For $t \neq \infty$ the fibre X_t of $\tilde{X}$ over $t \in \mathbb{P}^1$ is isomorphic to X, while $X_\infty = X'_\infty \cup X''_\infty$, where $X''_\infty = X, X'_\infty \cap X''_\infty = Y$, and X'_∞ is a $\mathbb{P}^1$-bundle over Y. Also $\tilde{Y} = Y \times \mathbb{P}^1$ injects into $\tilde{X}$, and does not meet X''_∞.

We let $E^\bullet$ denote the complex $I_{\tilde{Y}} \cdot E(X''_\infty) \hookrightarrow E(X''_\infty)$ of bundles on $\tilde{X}$. We endow $E(X''_\infty)$ and $I_{\tilde{Y}}$ with hermitian metrics such that in an open neightborhood U of X''_∞ the metric of $I_{\tilde{Y}}$ coincides with that on $O_{\tilde{X}}$. This also gives a metric on $I_{\tilde{Y}} \cdot E(X''_\infty)$. Furthermore we denote by $F^\bullet$ the complex $f_*(I_Y \cdot E) \hookrightarrow f_*(E)$ on B. Then on $B \times \mathbb{P}^1$ we have an augmentation $v\colon F^\bullet \to f_*(E^\bullet) \to f_*(E^\bullet \otimes \wedge^* \Omega^{0,1}_{X/B})$, which vanishes on X''_∞. Also $v_\infty\colon F^\bullet_\infty \to f_*(E^\bullet/X'_\infty)$ denotes the augmentation at infinity.

We intend to show that $\lim\limits_{t \to \infty} \tilde{\mathrm{ch}}\left(\mathrm{cone}(v_t)\right) = \tilde{\mathrm{ch}}\left(\mathrm{cone}(v_\infty)\right)$. Here the singularities around ∂X_∞ may cause difficulties. Therefore we define a second family of bundles

$$E^{\bullet,*} = \left(E(X''_\infty) \xrightarrow{\sim} E(X''_\infty)\right), \quad F^{\bullet,*} = \left((0) \to (0)\right)$$

Obviously the conclusion of Theorem 6.3 holds for $E^{\bullet,*}$ as everyting is split and all $\tilde{\mathrm{ch}}$-classes vanish. Moreover on U $E^\bullet$ and $E^{\bullet,*}$ are isomorphic, and the augmentation v has small norm. We thus may expect that the singularities around ∂X_∞ will cancel out if we consider the difference $\tilde{\mathrm{ch}}_t - \tilde{\mathrm{ch}}^*_t$.

The metrics on X_t will tend to become singular near ∂X_∞. To remedy this we introduce a metric on $\Omega_{\tilde{X}/B \times \mathbb{P}^1}(d\log \infty)$ and its dual, the logarithmic relative tangent-bundle. We do this by changing the original Kähler-metric in an open neighborhood of ∂X_∞, contained in U. This induces hermitian metrics on all X_t as well as $X'_\infty - \partial X_\infty$ and $X''_\infty - \partial X_\infty$, which however may not be Kähler. Locally $\tilde{X}$ is isomorphic to the product of $Y - \partial X_\infty$

with the product of two unit disks $\{|z| < 1, |w| < 1\}$, and the projection to $\mathbb{P}^1$ is given by $\frac{1}{t} = xw$. If we use $\log(z)$ and $\log(w)$ instead, identifying the punctured unit disc via $\exp(2\pi i \tau)$ with $\mathcal{H}/\mathbb{Z}(\mathcal{H} = $ upper half-plane), then the metric on X_t looks like a small perturbation of a translation-invariant metric on $Y \times \mathbb{C}$, and similar for $X'_\infty - \partial X_\infty$ and $X''_\infty - \partial X_\infty$. We thus find a good finite system of coordinate charts, uniform for each X_t, and can define Sobolev-norms with them: For a positive integer $s \geq 0$ H_s is the completion of C_0^∞ under square-integration of all derivatives up to order s, and H_{-s} its dual. As the metric is complete. (∂X_∞ has infinite distance from any point in $X_\infty - \partial X_\infty$) one checks as before that C_0^∞ is also dense in H_{-s}, that there is a uniform Garding inequality for the $\bar{\partial}$-Laplacian Δ, and that Δ extends to a self adjoint operator whose domain of definition is the image of $H_2 \to H_0$. In short we have smoothed out all singularities as $t \to \infty$. However the price we pay for this is that the volume of X_t approaches infinity as $t \to \infty$, so that estimates in sup-norm do not imply L^2-estimates as easily as before. Especially an integral operator may not be trace-class anymore.

Next we address the fact that our metric is no more Kähler:
The Kähler-condition has been used to compute the singular terms in various asymptotic expansions. In our case the X_t are only non-Kähler in places where the complex is split, so that all the asymptotics vanish there anyway.

In more detail one first derives that there is a super- Laplacian $\widetilde{\Delta}_u$, limit of ordinary Laplacians as we blow up the metric on B. The only new feature is that some terms in it may have Grassmann-degree > 2 and thus scale with negative powers of u.

Then we define the $\widetilde{\text{ch}}$-class via the regularized integral
$\int\limits_0^\infty \text{tr}_s \left(\widetilde{N} \cdot \exp(\widetilde{\Delta}_u) \right) \frac{du}{u}$. Under an infinitesimal change of metric its derivative
is the Schwartz-limit
$\lim\limits_{u \to 0} \text{tr}_s \left(Q \cdot \exp(-\widetilde{\Delta}_u) \right)$, Q the hermitian operator describing the change. If Q has support in U this vanishes because of the splitting, except for terms related to the augmentation v. If we replace v by $\mu^{\frac{1}{2}} \cdot v$, with a parameter μ between 0 and 1, the derivation $\mu \cdot \frac{\partial}{\partial \mu}$ of this class is equal to the Schwartz-limit $\lim\limits_{U \to 0}'$ of the derivative (described by Q) of $\text{tr}_s \left(N_F \cdot \exp(-\widetilde{\Delta}_u) \right)$, where N_F denotes the number-operator which is identically 1 on $F^\bullet$ and vanishes on $E^\bullet$. Integrating we obtain that the $\mu \frac{\partial}{\partial \mu}$-derivative of the effect of change of metric is given by the difference in Schwartz-limits

$$\lim\limits_{u \to 0}' tr_s \left(N_F \cdot \exp(-\widetilde{\Delta}_u) \right),$$

taken once for the original Kähler-metric and once for the logarithmic non-Kähler-metric. But for the original metric this term does not depend on μ and is equal to the chern-character of $F^\bullet$, while for the new metric the term converges nicely as $t \to \infty$. (See Lemma 6.8 below) This also holds if we integrate against $d\mu/\mu$. So all in all the change in metrics contributes a correction-term which converges for $t \to \infty$ to the corresponding term at infinity, and thus does not affect the conclusion of the theorem.

Finally we have to consider $\tilde{\mathrm{ch}}$-classes on X'_∞ and X''_∞. Let us concentrate on X'_∞, as on the other component everything will cancel out anyway at the end.

As before Δ defines a heat-kernel $\exp(-u\Delta)$ on $X'_\infty - \partial X_\infty$, and similar for the super-analogues. However because of the infinite volume this kernel is no more of trace-class. What we shall do is to identify $E^\bullet$ and $E^{\bullet,*}$ on $U \cap X'_\infty$, and consider the difference of heat-kernels $\exp(-u\Delta)$ and $\exp(-u\Delta^*)$. We shall see that it is in fact integrable, and so we can still define the difference $\tilde{\mathrm{ch}}_\infty - \tilde{\mathrm{ch}}^*_\infty$, although the individual terms do not make sense.

With this notation we shall show that $\lim_{t \to \infty}(\tilde{\mathrm{ch}}_t - \tilde{\mathrm{ch}}^*_t) = \tilde{\mathrm{ch}}_\infty - \tilde{\mathrm{ch}}^*_\infty$. It remains to relate the right-hand side to what it was claimed to be in Theorem 6.3. For this we repeat the same procedure with the embedding $Y \hookrightarrow X'_\infty$ and the original Kähler-metric on X'_∞. For this deformation all fibres X_t are isomorphic to X'_∞, and one sees that Theorem 6.3 holds for it. But our construction gives the same (logarithmic) $\tilde{\mathrm{ch}}_\infty - \tilde{\mathrm{ch}}^*_\infty$ as for the deformation associated to $Y \hookrightarrow X$, so that finally adding up all the pieces the assertion of Theorem 6.3 has been shown.

Now we need some technical results. We formulate them for $E^\bullet$, but they also hold for $E^{\bullet,*}$. In the following we need operators on $F^\bullet \oplus f_*(E^\bullet \otimes \wedge^* \Omega_X^{0,1})$. We usually ignore the $F^\bullet$-part. In fact we usually distinguish between local behavior near ∂X and away from it. Then Y behaves as if it were in the second category. For example cutoffs near ∂X operate trivially on Y. Let $D = \bar{\partial} + \bar{\partial}^* + v + v^*$ denote the Dirac-operator, and $\Delta = D^2$ its square.

Consider $X_\infty = X'_\infty \cup X''_\infty \cdot \partial X_\infty = X'_\infty \cap X''_\infty$ has real codimension 2 in X_∞.

In the degeneration process the covariant derivatives ∇ behave well, as $\nabla =$ hermitian connection. Similar for volume forms. We need

Lemma 6.6. *If $\phi, D\phi, \in L^2 \cap C^\infty(X'_\infty - \partial X'_\infty)$, $\psi \in C^\infty(X'_\infty)$ implies that*

$$\int_{X_\infty} <\phi, D\psi> = \int_{X_\infty} <D\phi, \psi>.$$

Proof. Elementary using cutoffs.

As a corollary, using the above formula we get :

Lemma 6.7. *Assume* $\phi \in C^\infty(X'_\infty - \partial X'_\infty)$, ϕ, $D\phi$ *and* $\Delta\phi$ *are in* L^2. *Then* $\Delta\phi$ *represents the distribution* $\Delta\phi$ *on* X'_∞.

Of course these assertions also hold for X''_∞.

Lemma 6.8. *Let* $K_t(u, x, y)$ *denote the heat kernel* $e^{-u\Delta}$ *on* X_t. *We claim that uniform on any compactum of* $]0, \infty[\times(X_\infty - \partial X_\infty)^2$, *the* $K_t(x, y)$ *converge to the heat kernel* $K_\infty(x, y)$ *on* $X'_\infty \amalg X''_\infty$.

Proof. By standard arguments, for any sequence of $t_n \to \infty$ we can find a subsequence such that $K_{t_n}(u, x, y)$ converge in the C^∞-topology, uniformly on compacta in $]0, \infty[\times(X_\infty - \partial X_\infty)^2$. We have to show that this limit $K(u)$ is necessarily $K_\infty(u)$.

First by passing to $\lim_{t\to\infty}$ we have (as C^∞-functions on $X_\infty - \partial X_\infty$)

$$(\frac{\partial}{\partial u} + \Delta)K(u) = 0.$$

As

$$\|K_{t_n}(u)(\phi) - \phi\|_{L^2} \leq u\|\Delta\phi\|^2_{L^2}$$

$(1 - e^{-\lambda u} \leq \lambda u$ for all $\lambda)$ for $\phi \in C_0^\infty(X_{t_n})$, it follows that

$$\lim_{u\to 0} K(u)\phi = \phi$$

for $\phi \in C_0^\infty(X_\infty - \partial X_\infty)$. Also for such a ϕ we have that

$$\|DK(u)\phi\|_{L^2} \text{ and } \|\Delta K(u)\phi\|_{L^2}$$

are finite (pass to the limit) so by 6.7

$$(\Delta + \frac{\partial}{\partial u})K(u) = 0$$

as distributions. This implies that $K(u)(\phi) = e^{-u\Delta}(\phi)$.

Proposition 6.9.

   There exists a positive $\lambda > 0$ *such that uniformly in each fibre* $f^{-1}(b)$, b *in* B, *and for each* $t \in \mathbb{P}^1$, *the Laplacian on* cone(v_t) *has all eigenvalues* $\geq \lambda$. *The same is true for* cone(v_∞).

Proof. Let us start with the assertion about the v_t. There can only be a problem as $t \to \infty$. So if the assertion is wrong we can find a sequence $t_n \to \infty$, and eigenfunctions f_n on $X_n = X_{t_n}$ with eigenvalues $\lambda_n \to 0$. Suppose $\|f_n\| = 1$. Then by Sobolev estimates the f_n are uniformly bounded in the

C^∞-topology, and we can choose a subsequence which converges on each compactum in $\tilde{X} - \partial X_\infty$, and on B. The limit must be annihilated by $\Delta_{\bar\partial}$. We claim that it must vanish identically. If this holds true then the f_n must concentrate their mass around X''_∞ where $E^\bullet$ is metrically split. Cutting them off we can find a new sequence f_n with $\|f_n\| = 1$, such that the f_n are supported in $U \cap X_n$, and $\|\Delta_{\bar\partial} f_n\| < 1$. But as $E^\bullet$ is split in U, there $\Delta_{\bar\partial}$ is the sum of 1 and a positive operator. Thus $< \Delta_{\bar\partial} f_n, f_n > \geq < f_n, f_n > = 1$, a contradiction. It now suffices to treat X'_∞. Suppose again that we have a sequence of eigenfunctions f_n with $\|f_n\| = 1$, such that $\Delta_{\bar\partial} f_n = \lambda_n \cdot f_n$, and $\lambda_n \to 0$. Cutting off we see as before that the square-integrals of f_n over $U \cap X_n$ converge to zero. Passing to a subsequence we may assume that the f_n converge to an f with $\Delta_{\bar\partial}(f) = 0$, $\|f\| = 1$, and f vanishes identically on U. Then in the original Kähler-metric on X'_∞ f defines also a harmonic form, so there cone(v_∞) has non-trivial cohomology. However this contradicts our choice of $F^\bullet$.

Theorem 6.10. *On compacta in $X_\infty - \partial X_\infty$ we furthermore have uniformly*

(1) *for $u \to 0$ the asymptotic expansion of $K_t(u, x, y)$ converges to $K_\infty(u, x, y)$ uniformly.*
(2) *for $u \to \infty$ $K_t(u, x, y)$ decays uniformly exponentially.*

Proof. (1) follows from the local nature of the expansion. (2) follows as the Sobolov-norms decay exponentially, because all eigenvalues of Δ are $\geq \lambda > 0$.

In short we have complete control on $X - \partial X$, uniform on any compactum.

Using perturbation expansions these assertions also hold for the super-connection $\tilde\nabla_u^2$ as well as operators $\tilde\nabla_u^2 + \alpha \cdot \tilde{N}$, α a small parameter. Moreover the result is C^∞ in α. However for small u we have to be careful as $\tilde\nabla_u^2$ and $\tilde{N}$ may contain Grassmann-terms which scale with negative powers of u as $u \to 0$. One still derives estimates in the sense of asymptotic expansions: A function $f(u) = \sum_{i=-n}^{-1} a_i u^i + g(u)$ ($g(u)$ regular at $u = 0$) is small in this sense if all a_i as well as g is small. Slightly stronger in all cases of interest for us it will be true that $u^n \cdot f(u)$ as well as all its u-derivatives are small. To prove these estimates one uses the usual technique of blowing-up the metric on B and comparing to the tangent-space.

It remains to consider what happens in the open neighborhood U of X''_∞. There we can identify the two complexes $E^\bullet$ and $E^{\bullet,*}$. Then the Laplacian Δ on $E^\bullet$ is equal to the sum of a local operator Δ^* which coincides with the Laplacian for $E^{\bullet,*}$, and an integral operator Δ_{gl} which comes from the augmentation v. The latter is determined by global sections of $F^\bullet$. Furthermore it is of trace-class norm $O(\epsilon)$ if we restrict to an ϵ-neighborhood of X''_∞, in the original metric. This also holds for superconnections if interpreted accordingly: Here the difference $\widetilde{\nabla}_u^2 - \widetilde{\nabla}_u^{*2}$ is a sum of $u\cdot$ (an integral operator of Grassmann-degree zero) and an integral operator of Grassmann-degree ≥ 1. The first term is as before, while the second one is made up from covariant derivatives (in B-direction) of elements of $F^\bullet$. These vanish on ∂X_∞ (because we use logarithmic metrics), and thus again the operator has trace-class norm $O(\epsilon)$.

Now we come to the proof of Theorem 6.3. Let φ_ϵ denote a cutoff-function with support in an 2ϵ-neighborhood of X''_∞, which is $\equiv 1$ at distance $\leq \epsilon$ from ∂X_∞. Also φ_ϵ should have support in U and the C^∞ norm of φ_ϵ (in the logarithmic norm) should be uniformly bounded. As usual this also should mean that φ_ϵ acts as zero on $F^\bullet$. Then as on U we can identify $E^\bullet$ and $E^{\bullet,*}$, it makes sense to consider

$$L_t(u) = \varphi_\epsilon \big(K_t(u) - K_t^*(u)\big)\varphi_\epsilon,$$

as an operator on $f_*\big(E^{\bullet,*} \otimes \wedge^*\Omega^{0,1}_{X_t/B}\big)$. Then

$$(\Delta^* + \frac{\partial}{\partial u}) L_t(u) = \Delta^*\varphi_\epsilon \big(K_t(u) - K_t^*(u)\big)\varphi_\epsilon$$
$$- \varphi_\epsilon\big(\Delta\,K_t(u) - \Delta^*\,K_t^*(u)\big)\varphi_\epsilon$$
$$= [\Delta^*, \varphi_\epsilon]\,\varphi_{2\epsilon}\big(K_t(u) - K_t^*(u)\big)\varphi_\epsilon$$
$$- \varphi_\epsilon\,\Delta_{gl}\,K_t(u)\varphi_\epsilon$$

where $\varphi_\epsilon\,\Delta_{gl}$ has trace-class norm $0(\epsilon)$.

As $\lim\limits_{u \to 0} L_t(u) = 0$, we obtain

$$L_t(u) = u \int\limits_{a+b=1} K_t^*(au)[\Delta^*, \varphi_\epsilon]\varphi_{2\epsilon}\big(K_t(bu) - K_t^*(bu)\big)\varphi_\epsilon\,da$$
$$- u \int\limits_{a+b=1} K_t^*(au)\,\varphi_\epsilon\Delta_{gl}\,K_t(bu)\,\varphi_\epsilon\,da$$

The second term is of trace-class norm $O(\epsilon\,u\,e^{-\lambda u})$ uniformly in t for u big. For small u it is $O(\epsilon)$ in the asymptotic sense, that is all terms in the sympototic expansion will be of trace-class norm $O(\epsilon)$. For the first integral we apply the same procedure from the right, i.e. apply $(\Delta^* + \frac{\partial}{\partial u})$ from the other side, to obtain

$$L_t(u) =$$

$$u^2 \int\limits_{a+b+c=1} \big\{ K_t^*(au)[\Delta^*, \varphi_\epsilon]\varphi_{2\epsilon}\big(K_t(bu) - K_t^*(bu)\big) \ \varphi_{2\epsilon}[\varphi_\epsilon, \Delta^*]K_t^*(cu)\big\} \, dadb$$

$$+ O(\epsilon \, u^2 e^{-\lambda u})$$

We are interested in $\text{tr}_s \left(\widetilde{N} \, L_t(u) \right)$, which now turns out to be

$$\text{tr}_s \left(\widetilde{N} \cdot L_t(u) \right) =$$

$$u^2 \int\limits_{a+b+c=1} \text{tr}_s \big(K_t^*(cu)\widetilde{N} \, K_t^*(au)\big)[\Delta^*, \varphi_\epsilon]\varphi_{2\epsilon}\big(K_t(bu) - K_t^*(bu)\big)\varphi_{2\epsilon}[\varphi_\epsilon, \Delta^*]\big) dadb$$

$$+ O(u^2 \epsilon \, e^{-\lambda u})$$

$$= -u \int\limits_{a+b=1} \text{tr}_s \big(\frac{\partial}{\partial\alpha_{\alpha=0}} e^{-au\Delta^* + \alpha \widetilde{N}}[\Delta^*, \varphi_\epsilon]\varphi_{2\epsilon}\big(K_t(bu) - K_t^*(bu)\big)\varphi_{2\epsilon}[\varphi_\epsilon, \Delta^*]\big) da$$

$$+ O(u^2 \epsilon \, e^{-\lambda u})$$

This also holds for superconnections (pass to the limit), where $a\,u\,\Delta^*$ has to be replaced by $a\widetilde{\nabla}_u^{*2}$, etc. Note that the trace is now an integral over $\text{supp}(d\varphi_\epsilon)^2$. Thus for fixed ϵ we are in a compact set of $\widetilde{X} - \partial X_\infty$, so everything decays nicely for $u \to \infty$, has good asymptotic expansions, etc. We may thus form the regularized integral

$$\int\limits_0^\infty \text{tr}_s \left(\widetilde{N} L_t(u) \right) \frac{du}{u},$$

and it follows that up to a term $O(\epsilon)$ this converges for $t \to \infty$ to the corresponding integral at ∞. Moreover these integrals are the contribution to $\widetilde{\text{ch}}_t - \widetilde{\text{ch}}_t^*$ which comes from integrating the appropriate kernel over $\text{supp}(\varphi_\epsilon)$. For the rest we have convergence as before, so finally

$$\lim_{t \to \infty} \widetilde{\text{ch}}_t - \widetilde{\text{ch}}_t^* = \widetilde{\text{ch}}_\infty - \widetilde{\text{ch}}_\infty^* \, .$$

LECTURE 7. THE THEOREM OF BISMUT–VASSEROT

Suppose X is a projective complex algebraic manifold, L an ample line-bundle on X, and E a hermitian vectorbundle. Then we know that for big q $E \otimes L^{\otimes q}$ has only cohomology in degree zero. Endow $\Gamma(X, E \otimes L^{\otimes q})$ with its L^2-metric. If everything is defined over the integers of a numberfield we obtain a lattice of integral points in it, and for arithmetic applications one often needs to know its covolume. This is computed by the arithmetic Riemann-Roch theorem, up to the $\widetilde{\mathrm{ch}}$-class of the cone of v: $\Gamma(X, E \otimes L^{\otimes q}) \to \overline{\Omega}^*$-complex, which in turn is the regularized integral

$$\widetilde{\mathrm{ch}}(v_q) = -\frac{1}{2} \int_0^\infty \mathrm{tr}_s \left(N \exp(-u\Delta_q) \right) \frac{du}{u}.$$

Here N is the number-operator on the $\overline{\Omega}^*$-complex. In [BV] Bismut and Vasserot give bounds for this. The main strategy is as follows:
Split the integral in two parts, one where qu is bounded and the other where qu approaches ∞. For the first part we obtain asymptotic expansions, uniform in q. In fact for this we need no ampleness. For the second we have to use that the eigenvalues of Δ_q are sufficiently positive, so that it decays fast enough. Although this is not necessary we assume for simplicity that X is Kähler. First we treat small times:
Choose a point $x \in X$. We make a local computation around x, but as usual everything will be uniform and C^∞ in x. Choose holomorphic coordinates $z_1, \ldots, z_n$ around x $(n = \dim_{\mathbb{C}}(X))$ such that the metric is given by a Kähler- potential

$$\varphi_X = \sum_{j=1}^n |z_j|^2 + (\text{order} \geq 4)$$

Also choose an ON-frame for E near x. Furthermore we can find a local holomorphic generator f of L for which

$$\log \|f\|^2 = -\sum_{j,k} a_{jk} z_j \bar{z}_k + (\text{order} \geq 3)$$

The a_{jk} define a hermitian endomorphism $N(L)$ of $T^{1,0}_{X,x}$, independent of the coordinates. Also the curvature $R(L)$ of L at x is equal to

$$R(L) = \sum_{j,k} a_{jk} dz_j \wedge d\bar{z}_k + (\text{order} \geq 1)$$

We can choose the z_j in such a way that $N(L)$ is diagonal, i.e.

$$R(L)(x) = \sum_j \lambda_j dz_j \wedge d\bar{z}_j.$$

This simplifies the linear algebra below.

Now denote by $K_q(u, x, y) \cdot$ (volume form) the heat-kernel $e^{-u \Delta_q}$.

Theorem 7.1.

    a) $(2\pi u)^n q^{-n} K_q(q^{-1}u, x, x) = L_q(u, x) \in \text{End}\left(E \otimes S(X)\right)(x)$ is C^∞
       for

       $u \in [0, \infty)$, $x \in X$, and has a limit $L_\infty(u, x)$ as $q \to \infty$

    b) On any compactum, $L_q(u, x) - L_\infty(u, x)$ is $O(q^{-\frac{1}{2}})$ in the C^∞-
       topology (in the asymptotic sense as $u \to 0$)

    c) $L_\infty(u, x) = \det\left(\frac{uN(L)}{1 - e^{-uN(L)}}\right) e^{-uN(L)}$ where $\mathcal{N}(L)$ is the derivation
       on $\Lambda^* \Omega_X^{0,1}$ which sends $d\bar{z}_j$ to $\lambda_j \cdot d\bar{z}_j$

    d) $\text{tr}_s\left(N \cdot L_\infty(u, x)\right) = \det(u\, N(L)) \cdot \text{tr}\left(\frac{1}{1 - e^{u N(L)}}\right) \cdot rk(E)$

Proof: We have already trivialized the bundles near x. In these trivializations the Christoffel-symbols $\Gamma(X)$ and $\Gamma(L)$ vanish at the origin. Now we scale the coordinates by a factor $(uq^{-1})^{\frac{1}{2}}$. Then all the curvature terms get a factor uq^{-1}, so only the curvature of $L^{\otimes q}$ survives in the limit $q \to \infty$. That is by the Lichnerowicz-formula in the new coordinates the Laplacian is given by (note that $\Delta = \frac{1}{2} D^2$)

$$2uq^{-1} \cdot \Delta_q = -\sum_j (\partial_j - \frac{u}{2} \cdot \sum_k R_{jk}(L)x_k)^2 + u\, R(L)^a + O((uq^{-1})^{\frac{1}{2}})$$

So by the usual perturbation theory of lecture 3 we obtain a) and b), with the limit determined by the limiting Laplacian above, without the error term. There the leading term is give by Getzler's computation [Ge] as

$$\prod_j \frac{u\lambda_j}{e^{\frac{u}{2}\lambda_j} - e^{-\frac{u}{2}\lambda_j}} \cdot \exp\left(-\frac{u}{2} \cdot \sum_j \lambda_j (dz_j \wedge d\bar{z}_j)^a\right)$$

But $\sum_j \lambda_j (dz_j \wedge d\bar{z}_j)^a = -\text{tr}\left(N(L)\right) + 2 \cdot \mathcal{N}(L)$, and we derive c).

For d) we need to do a linear algebra computation:
$\Lambda^* \Omega_{X,x}^{0,1}$ has a basis $d\bar{z}^I$, $I \subseteq \{1, \ldots, n\}$ any subset. On this basis N acts as $|I|$, and $\mathcal{N}(L)$ as $\sum_{j \in I} \lambda_j$. Thus we need to compute

$$\sum_I (-1)^{|I|} \cdot |I| \cdot \exp(-u \sum_{j \in I} \lambda_j)$$

$$= \frac{d}{dt}_{t=1} \prod_j (1 - t \exp(-u\lambda_j))$$

$$\prod_j (1 - \exp(-u\lambda_j)) \cdot \sum_j \frac{-\exp(-u\,\lambda_j)}{1 - \exp(-u\,\lambda_j)},$$

which gives the desired result.

Large times.

Next we consider what happens for uq^{-1} big. Let μ_q denote the infimum of the eigenvalues of $q^{-1}\Delta_q$ on forms of degree ≥ 1. μ_q is positive for big q, as then the complex is acyclic in positive degrees.

Lemma 7.2.

    a) *Assume $\mu \in \mathbb{R}$ is less than the eigenvalues λ_i of $N(L)(x)$, for any $x \in X$. Then for q big enough $\mu_q > \mu$.*

    b) *In any case there exists a constant c such that for big q $\mu_q \geq \exp(-c \cdot q)$*

Proof: Assertion a) holds for the limiting Laplacian

$$\frac{1}{2}\Big(-\sum_j \big(\partial_j - \frac{1}{2}\sum_k R_{jk}(L)x_k\big)^2 + R(L)^a\Big),$$

since it respects the basis $d\bar{z}^I$, is ≥ 0 on scalars (=0-forms), and its action on $d\bar{z}^I$ differs from that on scalars by the positive constant $\sum_{j \in I} \lambda_j > \mu$, if $I \neq \emptyset$. (The curvature term $R(L)^a$ itself is not positive, but the sum with the first term is.) It follows by approximation that there is a neighborhood U_x of x such that for any section e of $E \otimes L^{\otimes q} \otimes S(X)$, of positive degree and with support in U_x, we have

$$< \Delta_q e, e > \geq q\mu < e, e >$$

or equivalently

$$\|(\bar{\partial} + \bar{\partial}^*)e\| \geq \sqrt{q\mu} \cdot \|e\|$$

Choose finitely many C^∞-functions φ_i such that each φ_i has support in some U_x, and such that $\sum_i \varphi_i^2 = 1$. Then $[\bar{\partial} + \bar{\partial}^*, \varphi_i]$ is bounded in L^∞, independently of q. Thus if e is a global section of $E \otimes L^{\otimes q} \otimes S(X)$, which is an eigenvector for $\bar{\partial} + \bar{\partial}^*$ with eigenvalue ξ, then

$$
\begin{aligned}
\xi \cdot <e, e> &= \sum_i <\varphi_i(\bar{\partial} + \bar{\partial}^*)e, \varphi_i(\bar{\partial} + \bar{\partial}^*)e> \\
&= \sum_i <(\bar{\partial} + \bar{\partial}^*)\varphi_i e, (\bar{\partial} + \bar{\partial}^*)\varphi_i e> + O(\sqrt{\xi}\,\|e\|^2) \\
&\geq \mu q \cdot \sum_i <\varphi_i e, \varphi_i e> + O(\sqrt{\xi}\,\|e\|^2) \\
&= \mu q \cdot \|e\|^2 + O(\sqrt{\xi}\|e\|^2,
\end{aligned}
$$

so $\xi \geq \mu q + O(\sqrt{\xi})$.

Making μ a little bit smaller and choosing q big gives the result.

For b) we observe (following [BV]) that the minimal eigenvalue $q\mu_q$ of Δ_q in positive degrees occurs in its restriction to $\overline{\mathrm{Im}(\bar{\partial})}$. Now we easily obtain the following description of μ_q:

μ_q is the biggest number such that for any $h \in \overline{\mathrm{Im}(\bar{\partial})}$, there exists a g with $\|g\|^2 \leq (q\mu_q)^{-1} \cdot \|h\|^2$ and $\bar{\partial}g = h$: For one direction decompose into eigenspaces, and for the other decompose g into $\mathrm{Ker}(\bar{\partial}) \oplus \mathrm{Ker}(\bar{\partial})^\perp$.

Now for any two metrics on L there exists a constant $c > 0$ such that the metrics on $E \otimes L^{\otimes q} \otimes S(X)$ are mutually bounded by $\exp(\frac{c}{2}q)$. From the above description it follows that the ratio of the minimal q eigenvalue $q\mu_q$ is bounded by $\exp(c \cdot q)$. By a) it is bounded below if we choose a metric with positive curvature (which exists as L is ample), so in general we have a lower bound $\exp(-c \cdot q)$.

Application to analytic torsion.

Finally want to compute the regularized integral

$$
\widetilde{\mathrm{ch}}_q = -\frac{1}{2} \int_0^\infty \mathrm{tr}_s(Ne^{-u\Delta_q}) \frac{du}{u}.
$$

We fix a big u_0, and split into $u \leq u_0/q$ and $u \geq u_0/q$. For the first integral we would like to introduce the new variable qu. However as we have a regularized integral one checks that for such a substitution

$$\int\limits_0^\infty f(qu)\,\frac{du}{u} \;=\; \int\limits_0^\infty f(u)\frac{du}{u} - \log(q)\lim_{u\to 0}{}' f(u),$$

where $\lim'$ denotes the Schwartz-limit.

Thus

$$-\frac{1}{2}\int\limits_0^{u_0/q} \mathrm{tr}_s(N e^{-u\Delta_q})\,\frac{du}{u}$$

is equal to the X-integral of

$$-\frac{1}{2}\int\limits_0^{u_0/q} \left(\frac{1}{2\pi u}\right)^n \mathrm{tr}_s\left(N\, L_q(qu, x)\right)\frac{du}{u}$$

$$= -\frac{1}{2}\left(\frac{q}{2\pi}\right)^n \left(\int\limits_0^{u_0} \mathrm{tr}_s\left(u^{-n} N\, L_q(u, x)\right)\frac{du}{u} - \log(q)\lim_{u\to 0}{}'\,\mathrm{tr}_s\left(u^{-n}\cdot N\, L_q(u, x)\right)\right)$$

Up to an error of size $o(q^n)$ we may replace L_q by L_∞. It already follows that the total contribution is $O\!\left(q^n\cdot\log(q)\right)$.

If $N(L)$ is positive we can do better. We want to compute everything modulo $O(q^n)$. Then the integral disappears. Thus we have to consider

$$\frac{1}{2}\left(\frac{q}{2\pi}\right)^n \log(q)\cdot\lim_{u\to 0}{}'\det\left(N(L)\right)\mathrm{tr}\left(\frac{1}{1 - e^{uN(L)}}\right)\cdot rk(E)$$

Locally the trace is $\sum_j \frac{1}{1-e^{u\lambda_j}} = \sum_j \frac{-1}{u\lambda_j + \frac{u^2}{2}\lambda_j^2+\dots}$, which has constant term $\frac{n}{2}$. Thus we obtain

$$\frac{1}{4}\det\left(\frac{N(L)}{2\pi}\right) q^n\cdot\log(q^n)\cdot rk(E)$$

By Riemann-Roch the X-integral of $q^n\cdot\det\left(\frac{N(L)}{2\pi}\right)\cdot rk(E)$ is $\dim\Gamma\left(X, E\otimes L^{\otimes q}\right) + O(q^{n-1})$.

Finally there is the integral

$$-\frac{1}{2}\int\limits_{u_0/q}^\infty \mathrm{tr}_s(N\, e^{-u\Delta_q})\frac{du}{u} = -\frac{1}{2}\int\limits_{u_0}^\infty \mathrm{tr}_s(N e^{-\frac{u}{q}\Delta_q})\frac{du}{u}.$$

At $u = u_0$ $K_q(\frac{u}{q}, x, x)$ is uniformly $O(q^n)$.
Furthermore separately in each degree

$$\frac{\partial}{\partial u} \operatorname{tr} \left(K_q(\frac{u}{q}) \right) \leq -\mu_q \cdot \operatorname{tr} \left(K_q(\frac{u}{q}) \right).$$

It follows that the integrand is uniformly $O(q^n) \cdot e^{-\mu_q(u-u_0)}$, so the integral is $O(q^n \cdot |\log \mu_q|)$. For positive $N(L)$ this is $O(q^n)$, and in general $O(q^{n+1})$. Thus

Theorem 7.3.

  a) $\widetilde{\operatorname{ch}}(v_q) = O(q^{n+1})$
  b) *If $N(L)$ is positive*

$$\widetilde{\operatorname{ch}}(v_q) = \frac{1}{4} q^n \cdot \log(q^n) rk(E) \int_X \det \left(\frac{N(L)}{2\pi} \right) dx \; + \; O(q^n)$$

$$= \frac{1}{4} \log(q^n) \cdot \dim \Gamma(X, E \otimes L^{\otimes q}) + O(q^n)$$

REFERENCES

[ABP]　M. Atiyah, R. Bott, V.K. Patodi, *On the Heat Equation and the Index Theorem*, Invent. Math., **19** (1973), 279-330.

[BGS1]　J.M. Bismut, M. Gillet, C. Soulé, *Analytic Torsion and Holomorphic Determinant Bundles I, II, III*, Commun. Math. Phys., **115** (1988), 49-78, 79-126, 301-351.

[BGS2]　J.M. Bismut, M. Gillet, C. Soulé, *Bott Chern Currents and Complex Immersions*, Duke Math. J., **60** (1990), 255-284.

[BL]　J.M. Bismut, G. Lebeau, *Complex Immersions and Quillen Metrics*, Preprint Orsay, 1990, also C.R.A.S., **309** (I), 1989, 487-491.

[BV]　J.M. Bismut, E. Vasserot, *The Asymptotics of the Ray-Singer Analytic Torsion Associated with High Powers of a Positive Line Bundle*, Commun. Math. Phys., **125** (1989), 355-367.

[BS]　A. Borel, J.P. Serre, *Le théorème de Riemann Roch (d´ après Grothendieck)*, Bull. Soc. Math. de France, **86** (1958), 97-136.

[D]　P. Deligne, *Le déterminant de la cohomologie*, Contemporary Math., **67** (1987), 93-177.

[F]　G. Faltings, *Calculus on Arithmetic Surfaces*, Annals of Math., **119** (1984), 387-424.

[Ge]　E. Getzler, *A short proof of the Atiyah-Singer Theorem*, Topology, **25** (1986), 111-117.

[GS1]　H. Gillet, C. Soulé, *Arithmetic Intersection Theory*, Publ. Math. (IHES) **72** (1990), 93-174.

[GS2]　——————, *Characteristic Classes for Algebraic Vector Bundles with Hermitian Metric*, Annals of Math., **131** (1990), 163-203, 205-238.

[GS3]　——————, *Analytic Torsion and the Arithmetic Todd Genus*, Topology, **30** (1991), 21-54.

[GS4] —————————, *An arithmetic Riemann-Roch Theorem*, C.R.A.S., **309** (I) (1989), 929-932.

[Q1] D. Quillen, *Superconnections and the Chern Character*, Topology, **24** (1985), 89-95.

[Q2] —————————, *Determinants of Cauchy-Riemann Operators over a Riemann Surface*, Funct. Anal. Appl., **19** (1985), 31-34.

[Wa] F.W. Warner, *Foundations of Differentiable Manifolds and Lie Groups*, Scott, Foresman and Co., Glenview (1971).

[We] A. Weil, *Introduction a l' ètude des variétés kähleriennes*, Hermann, Paris,(1958).